PATIENT SAFETY
The PROACT® Root Cause Analysis Approach

T0330713

PATIENT SAFETY
The PROACT® Root Cause Analysis Approach

Robert J. Latino
Reliability Center, Inc.
Hopewell, Virginia

CRC Press
Taylor & Francis Group
Boca Raton London New York

CRC Press is an imprint of the
Taylor & Francis Group, an **informa** business

CRC Press
Taylor & Francis Group
6000 Broken Sound Parkway NW, Suite 300
Boca Raton, FL 33487-2742

First issued in paperback 2019

© 2009 by Taylor & Francis Group, LLC
CRC Press is an imprint of Taylor & Francis Group, an Informa business

No claim to original U.S. Government works

ISBN-13: 978-1-4200-8727-7 (hbk)
ISBN-13: 978-0-367-38670-2 (pbk)

Library of Congress Cataloging-in-Publication Data

Latino, Robert J.
    Patient safety : the PROACT root cause analysis approach / Robert J.Latino.
      p. ; cm.
    "A CRC title."
    Includes bibliographical references and index.
    ISBN 978-1-4200-8727-7 (hardcover : alk. paper) 1. Medical errors--Prevention. 2. Medical care--Quality control. 3. Health facilities--Quality control. I. Title.
    [DNLM: 1. Medical Errors--prevention & control. 2. Health
Facilities--organization & administration. 3. Outcome and Process Assessment (Health Care) 4. Safety Management--organization & administration. 5. Systems Analysis. W X 153 L357p 2008]

R729.8.L38 2008
610.68--dc22   2008022720

Visit the Taylor & Francis Web site at
http://www.taylorandfrancis.com

and the CRC Press Web site at
http://www.crcpress.com

Dedication

Contents

I   RECOGNIZING AND SUPPORTING THE VALUE OF PROACTION IN HEALTHCARE

II   EVENT PRIORITIZATION TECHNIQUES

III   UNDERSTANDING ROOT CAUSE ANALYSIS

IV   THE PROACT® ROOT CAUSE ANALYSIS (RCA) METHODOLOGY

V   A SOFTWARE TECHNOLOGY TO SUPPORT PROACTION

VI   FROM THEORY TO PRACTICAL APPLICATION

Author

Robert J. Latino is chief executive officer of Reliability Center, Inc. (RCI), a reliability consulting firm specializing in improving equipment, process, and human reliability. Mr. Latino received his bachelor's degree in business administration and management from Virginia Commonwealth University.

Latino has been facilitating Root Cause Analysis (RCA) and Failure Modes and Effects Analysis (FMEA) analyses with his clientele around the world for over 20 years and has taught over 10,000 students in the PROACT® methodology. He is co-author of numerous seminars and workshops on FMEA, Opportunity Analysis, and RCA as well as co-designer of the national award-winning PROACT Suite software package.

Latino is an author of *Root Cause Analysis: Improving Performance for Bottom Line Results*, 3rd ed. (Taylor & Francis [CRC Press], 2006), and was a co-author with Kenneth C. Latino of *Error Reduction in Healthcare: A Systems Approach to Improving Patient Safety* (AHA Press, 1999), and *The Handbook of Patient Safety Compliance: A Practical Guide for Health Care Organizations* (Jossey-Bass, 2005). His other publications include "Optimizing FMEA and RCA Efforts in Healthcare" (*ASHRM Journal,* 2004, 24(3) 21–28), and "MRIs: A Need for Risk Management and Patient Safety" (*Materials Management for Healthcare,* January 2006). He presented "Root Cause Analysis versus Shallow Cause Analysis: What's the Difference?" at the ASHRM 2005 National Conference. Latino has been published in numerous trade magazines on the topics of reliability, FMEA, opportunity analysis, and RCA and has been a frequent speaker on the topic at domestic and international trade conferences. Latino has applied the PROACT methodology to the field of terrorism and counterterrorism in "The Application of PROACT RCA to Terrorism/ Counter Terrorism Related Events" (*Intelligence and Security Informatics,* Kantor, P. et al., Eds., Springer, Atlanta, 2005, 579–589).

Mr. Latino is currently doing research on the healthcare culture in contrast to the industrial culture. This is an effort to make the appropriate modifications to the methodologies to successfully bridge the proactive technologies from industry to healthcare.

Foreword

It has been over nine years since the release of the Institute of Medicine report, which made clear what many healthcare professionals have long believed. A hospital can be a dangerous place. The now famous statistic released in the report that 44,000 and possibly as many as 98,000 Americans die each year of medical mistakes has been quoted many times over. While some might say that patient care is not much safer today than it was in 1999, I as a healthcare professional remain optimistic that patient safety will soon be recognized on the same level as medical science and technology.

It has been demonstrated over and over again that unintended medical errors are made by competent healthcare professionals, who are fallible individuals asked to work in often very dysfunctional environments and systems. The standard response and solution to medical error has been to blame the person or people at the sharp end; that is, those caregivers directly involved. In the best selling book *Internal Bleeding,* this traditional reactive approach is cited:

> We have become inured to and paralyzed by it, thinking of medical errors as the unavoidable collateral damage of a heroic, high-tech war we otherwise seem to be winning. It is as if we have spent the last 30 years building a really souped up sports car, but barely a dime on its bumpers, seatbelts, and airbags.

Given today's hospital environment, where fallible human beings come to work every day, the real question is not "why" did an unintended error occur but "how" did it happen?

Under pressure from state governments, Medicare, and large private insurers, many hospitals around the country are now agreeing not to charge when a so-called "Never Event" has occurred. Beginning in October 2008, Medicare began its no-payment program for never events. Hospitals have no choice under this program. They will not get paid. A spokesman for Medicare recently stated that the federal government expects to save $190,000,000 over five years on the nonpayment for never events.

The safety culture in healthcare has no choice but to change. Our organizations must move away from the traditional reactive response to a more proactive approach to ensure the delivery of safe medical care. Healthcare organizations, and those professionals and other workforce staff who practice within them, must be continuously mindful, respectful and focused on improving the reliability and safety of patient care. Unless continuous reinvestment is made to improve the systems and processes used by healthcare teams, there will be little accomplished to assure the patient of safe care.

We must continue in our efforts to explore, evaluate, and learn from high-reliability organizations (HROs). These HROs have a preoccupation with mistakes, remain always reluctant to simplify any event, are vigilant regarding the safety of their operations, will defer to the more experienced authority, and are committed in their ability to bounce back. For many years now, these organizations have been using root cause analysis (RCA), Failure Modes and Effectiveness Analysis (FMEA), and opportunity analysis (OA) with much success in improving the safety of their operations with identified return on the investment. As they have, our healthcare organizations must shift their focus from blaming the individual at the sharp end (those closest to the work or closest to the patients) to continuously identifying flows in the systems and processes of healthcare delivery that predispose clinicians to committing errors and cause or expose their patients to harm.

In this book, the authors give us the opportunity to learn how to use the tools of RCA, FMEA, and OA proactively in daily operations. Postmortems of medical errors often focus on the last few minutes before a harmful event occurs, and little attention is given to the contributing factors leading up to the event. By using these tools, an organization will better understand the plausible causes of harmful patient events, thus decreasing the risk of a recurrence. Of even more importance, these automated methods will give an organization the opportunity to analyze potentially harmful events before they occur. Standardized methodology, analysis, reporting, and tracking, and the ability to demonstrate that patient safety efforts are fiscally responsible, presents a good business case for the use of these methods.

While there have been many exciting advances in patient safety since the IOM report, we must continue to be ever mindful that healthcare is not only a high-tech industry but also one of high risk that must take into consideration at all times the patient, human factors, the high complexity of operations, and the many diversities of our healthcare teams.

Anne Flood, RN, MA
Director of RM/QM
Union Memorial Hospital
Baltimore, Maryland

Preface

The field of reliability engineering (RE) was developed initially in the aerospace field for the design, operation, and maintenance of aircraft. Reliability Center, Inc., was one of the first to pioneer these RE concepts from aerospace into heavy manufacturing in 1969. At that time, the Reliability Center was a research and development (R&D) arm of Allied Chemical Corporation. (Via many mergers and acquisitions, they are today Honeywell.) At this point in the history of our country, the manufacturing sector was doing very well, and profitability was outstanding.

Imagine introducing such proactive concepts as RE at a time of such economic prosperity. Cultural resistance was abounding, as no one could see why "proaction" was even needed. They were making lots of money. If something broke, they would just replace it.

As the world slowly became a global marketplace, competition became the impetus to lower unit costs. Manufacturers suddenly went from a paradigm of "if we make it, they will buy it" to "people are buying based on quality, reliability, and customer service." This is about the time the Japanese automakers took over the U.S. auto markets, and all were stunned.

I reflect on this cultural "experience," as this is what we see today. Healthcare is entrenched in centuries of tradition, which does not always support a proactive approach to the delivery of safe patient care. Medical errors are often thought of as the unavoidable parallel of a heroic, high-risk business. Sharing honest information about mistakes is not encouraged, as the present malpractice environment discourages it. Doctors fear that an admission of error or poor decisions will cost them their practice and livelihood. Nurses often feel that questioning a physician will be career limiting. Healthcare organizations pay more attention to the additional costs of care associated with harm, human error, technical failures, cost of litigation, and profit than they do to the existence of increased process complexities, information overload, and the importance of a team approach to healthcare delivery.

The predominance of errors created in healthcare today is largely the result of poor communications. While technical solutions exist, their initial costs are viewed as prohibitive. What often is not considered is their value over the long term and the lifecycle savings that will result.

Healthcare appears to have a difficult time in demonstrating that proactive activities increase patient safety while also improving the bottom line. Such value has been field proven in the manufacturing industries for at least the past 50 years.

Proactive tools such as root cause analysis (RCA), basic failure mode and effects analysis (FMEA), and opportunity analysis (OA) are affected by such restraining paradigms in healthcare. These tools seek to understand why things go wrong and to prevent mistakes from recurring. However, such tools are typically not high priorities and are viewed simply as *software* or *education* line items in the budget. They are not recognized for the dramatic potential impact they can have on patient safety and the bottom line in any organization. Unfortunately, they are often viewed simply as an additional cost and a burden to people already inundated with their daily reactive tasks. Such tools are competing for scarce healthcare dollars with other interests such as capital expenditures, reimbursement, equipment upgrades, building expansions, etc.

This book was written to address not only the proactive methodologies but also, more importantly, the organizational paradigms that must change in order to support and sustain such activities in the interest of patient safety. This text is unique because it is from the perspective of an entity *outside* of the healthcare industry, which has been through such a cultural transition before (reaction to proaction). I have spent 23 years in the reliability industry. I do not hold the traditional paradigms of those inside the healthcare system and therefore am neither biased nor constrained by them. I come to this more with the perspective of a patient who was oblivious to what goes on behind the healthcare "curtain" and then peeked behind the curtain and learned the realities.

This book challenges the conventional wisdom about what RCA is perceived to be. The reader will notice that we refer in the title of this text to the "PROACT approach." It is this approach that makes our definition of RCA unique. Most believe RCA to be simply a graphical tool to help depict a series of consequences. Our PRO-ACT approach includes tools such as Basic FMEA and OA, which help us to quantify and prioritize events that deserve the attention of RCA. It is for this reason we first address in this text events on which we should be doing RCA to maximize the effect on the patient.

Every one of us has a vested interest in optimizing patient safety at hospitals around the world, as we will probably all end up using them one day—especially as a baby boomer myself! When that time comes, we should have complete confidence in the care that is being provided to us and that we are safe when we are in a hospital. This is my small contribution to that vision.

Robert J. Latino
Chief Executive Officer
Reliability Center, Inc.

Acknowledgments

In my journey of learning the healthcare culture, I have had the pleasure of meeting new colleagues, but more importantly making new friends. The experience they have in healthcare covers the full spectrum. From lobbying for federal healthcare legislation to caring for patients in the ER, from dealing with malpractice claims to dealing with sentinel event investigations, they have excelled at them all.

I would like to acknowledge Fay A. Rozovsky, JD, MPH, DFASHRM for her support of our reliability concepts and technologies within the healthcare field over the past ten years. Her extensive experience in the fields of informed consent, disclosure, and risk management have been of extreme value in shaping my perspective of the current reality of healthcare.

I would like to thank my fellow colleagues in The Virtual Consulting Group (http://www.therozovskygroup.com/virtual/index.php), led by Fay Rozovsky. I have learned extensively from their collective expertise in the areas of credentialing, environment of care and safety, governance, The Joint Commission (TJC) readiness, medical group practice risk management, perinatal risk assessment, physician office practice risk management, and performance improvement. I am indebted for their willingness to share and grateful for their dedication to the healthcare field.

I would like to thank my friends at rL Solutions for having the courage to fully integrate their RMPro® incident management software with our PROACT® root cause analysis software. This combined package provides the breadth and depth of analysis that will be critical to prevent recurrence of undesirable outcomes, especially in light of the trend of nonpayment from entities such as Medicare and major insurance companies.

I would like to also acknowledge Patrice L. Spath, BA, RHIT, for her work in the quality field and her contribution and never-ending support of patient safety. A noted authority and author in this field, Patrice has been an admired confidant and influence on me during my journey into healthcare.

Next I would like to recognize Pamela L. Popp, MA, JD, FASHRM, CPHRM for sharing my vision of the potential impact of proaction on healthcare. Her extensive experience in dealing with claims and risk management allows her to see the worst of the problems in the industry and therefore to also recognize the solutions. Pamela has been instrumental in shaping my understanding of how the claims management field operates in theory and reality and how tools such as RCA can aid in the process.

I would also like to thank Robert M. Wachter, M.D., and Kaveh G. Shojania, M.D., for having the courage to publish their text *Internal Bleeding: The Truth Behind America's Terrifying Epidemic of Medical Mistakes* (New York: Rugged Land, 2004), which exposed many of the common deficiencies that exist in the current healthcare system. It was unique to see such a text written from "insiders" as opposed to "outsiders" looking in. Their text was another very big influence and supporter of my new perspective into the healthcare culture.

Last, but surely not least, I would like to acknowledge my star analysts who had the drive and initiative to demonstrate that the proactive tools described in this text work successfully in healthcare. They have proven to themselves, their organizations, and their patients that proaction saves lives and is good business. For that I would like to graciously thank the following for their impact on my life, but more importantly their impact on the lives of their patients.

MedStar Health System
> Anne Flood, RN, MA, Union Memorial Hospital, Baltimore, Maryland
> Arlene Feinstein, QM Coordinator, Union Memorial Hospital, Baltimore, Maryland
> Pam Grant, RM Coordinator, Union Memorial Hospital, Baltimore, Maryland

Bon Secours System
> Libby Enz, BSN, RN, ARM, CPHRM, CPHQ, Memorial Regional Medical Center, Richmond, Virginia
> Denise Allard, MHA, CPHRM, Community Hospital, Richmond, Virginia
> Steve Craig, MHA, Administration, Memorial Regional Medical Center, Richmond, Virginia

Loyola University Medical Center (LUMC)
> Monica Berry, BSN, JD, LLM, DFASHRM, CPHRM, Director, Patient Safety and Risk Management Loyola, LUMC, Oakbrook, Illinois

PeaceHealth
> Harold Peters, P. Eng., Process Design Manager, Portland, Oregon
> David Allison, Risk Manager, Sacred Heart Medical Center, Portland, Oregon

Fauquier Hospital, Inc.
> Shelby Forgacs, RN, Evaluator, Quality and Risk Management, Warrenton, Virginia
> Catherine Walsh, RN, BS, Director, Quality and Risk Management, Corporate Compliance Office, Warrenton, Virginia

I

Recognizing and Supporting the Value of Proaction in Healthcare

1 The Need for Reliability Tools in Healthcare

BACKGROUND INFORMATION

Based on our experiences, people who work in the healthcare field are one of the most committed groups to their profession with whom we have had the privilege to work. Many of the observations and candid conclusions in the pages that follow may give the reader the impression that this is not true. However, the comparisons presented between the industrial culture decades ago and healthcare today are meant to help the two communities come together in the interest of learning for the betterment of the entire healthcare industry and especially all of us as patients. The impediments will be explored that are imposed on the healthcare community by their unique form of communication.

In industry there is a well-known field called reliability engineering (RE). This field of engineering was pioneered in the heavy manufacturing industry by Charles J. Latino, who was the founder and director of the Reliability Center for Allied Chemical Corporation (now Honeywell) in 1972. RE is a field that focuses on equipment, process, and human reliability. Several methods and tools were modified for industry by this group, including root cause analysis (RCA), failure mode and effects analysis (FMEA), and a hybrid called opportunity analysis (OA). These three methods are singled out to be discussed in detail in later chapters, because they represent requirements under The Joint Commission (TJC) guidelines.

The perspectives and tools presented in this text have been field proven in the manufacturing industries for over 35 years. What this book hopes to accomplish is to convince the reader that reliability tools such as RCA, FMEA, and OA are *not* unique to any specific industry. Such reliability tools provide invaluable skills for human beings tasked with determining risk, opportunities, and why things go wrong.

The specific approach for the application of RCA, FMEA, and OA is referred to in the text as PROACT®.* While many approaches exist in the market, our tool of choice is PROACT, thus making it simply a brand name for our approach to identifying and solving problems.

The PROACT approach is used at such Fortune 500 companies as Lyondell-Citgo (oil refining), Weyerhaeuser (paper), Eastman Chemical (chemicals), Southern Companies (power), the U.S. Navy (defense), and many more industrial companies. In addition, healthcare systems such as MedStarHealth (Maryland), PeaceHealth (Oregon), and Loyola (Illinois) also use PROACT. The intent here is to demonstrate the range of applications. Whether it is problems making candy bars or problems aboard naval ships, the thought processes used to solve the issues are common to the

---

* PROACT is a registered trademark of Reliability Center, Inc.

1

analysts themselves. Solving problems in healthcare is no different. Humans are still tasked with figuring out what went wrong and need tools to accomplish the task.

THE SIGNALS WERE STACKING UP IN HEALTHCARE

In 1998, it became apparent that the healthcare industry was on the verge of a new frontier, that of patient safety. Several studies about the delivery of healthcare would be released to the press that would not be endearing to the medical community. It was also apparent that pending legislation would follow that would insist on efforts to reduce medical errors (or *medical harm*, as many would prefer to call it).

In short, the writing was on the wall; the healthcare industry would soon be "under the microscope" for deaths resulting from excessive medical errors. This tension in the industry climaxed with the release of the Institute of Medicine (IOM) report in 1999 indicating that between 44,000 and 98,000 people are killed per year by medical error.* This would surely create a public uproar, and it did, and a world-wide epidemic of medical error would be exposed.

As "outsiders" to this healthcare industry, such numbers are astounding. Deaths by medical error are the eighth leading cause of death for all Americans.† What is even more frustrating is finding out that the IOM report only considered "errors of commission." Errors of commission occur when someone takes an inappropriate action in handling a patient's care and the patient ends up worse than before as a result of the error.

The point here is that "errors of omission" were not in the report. Errors of omission are where someone *should have* taken action and did not. For instance, a patient comes into the emergency department (ED) with some symptoms that are not diagnosed properly by triage. As a result, the patient waits in the ED for an extended period of time and suffers a heart attack, a seizure, or any other adverse consequence due to not being seen in a timely manner. The number of people killed by errors of omission would likely be a significant multiple of those killed by acts of commission. Very few in the healthcare profession have openly contradicted the validity of the IOM report, and most believe that the statistics are grossly understated. What the IOM report did accomplish was a challenge to the idyllic belief that errors occur rarely; it therefore brought the fact that mistakes happen in healthcare to the forefront and gave the industry the opportunity to respond to the challenge.

THE PARADIGM OF PATIENT SAFETY

Looking at healthcare from an outsider's perspective is like going behind the curtain minutes before the opening of a Broadway play. The audience's perspective sees chaos that they would have never expected to see. This is where the traditional view of the hospital being a safe haven turns into the view of "avoiding getting sick or hurt at all costs so we do not have to run the risk of being admitted to the hospital."

---

* Kohn, Linda T., Corrigan, Janet M., and Donaldson, Molla S. (Eds.), *To Err is Human: Building a Safer Health System,* Institute of Medicine, Washington, DC, 2001.
† Centers for Disease Control and Prevention, National Center for Health Statistics, Deaths: Final Data for 1977, *National Vital Statistics Reports,* 47(19), 27, 1999.

Most Americans feel the traditional way. They see the hospital as a place where they can place their health and welfare in the competent professionals' hands, and the risk of a poor outcome is low. Even many healthcare professionals anonymously admit they have reservations about leaving their loved ones in a hospital. Even healthcare professionals indicate that patients must watch and question everything regarding a loved one's care. It is clear that education is necessary for both perspectives in order to gain a deserved respect for the other's position.

In a 2005 study conducted by VitalSmarts,* the following conclusions were published:

- Fifty-three percent of nurses and other clinical care providers expressed concerns about the competence of the peers they worked with.
- Eighty-one percent of physicians expressed concerns about a nurse's or other clinical-care provider's competence.
- Seventy-five percent of nurses and other clinical care providers expressed concerns about a peer's poor teamwork.

These few conclusions support our previous discussion, where most patients would not expect this type of discord "behind the curtain" of their hospitals.

Having attended many healthcare conferences, it was not rocket science to figure out that the buzzword in the healthcare industry was "patient safety" and to see its impact on patient outcomes and public opinion. Many very dedicated, talented, and esteemed speakers tell of their efforts to implement a culture of patient safety in their respective hospitals.

From the outsider's perspective, all we think about while listening to this is, "If patient safety is only now becoming the focus, what was it before?" From the patient's perspective, we would have thought patient safety was always the primary focus of healthcare!

Reflecting on experience from industry we ask, "What parallels could be drawn?" In industry, most every facility will have a safety department. It may be called something else, like Environmental, Health, and Safety (EH&S), but nonetheless it is charged with ensuring compliance with governmental safety regulations (i.e., EPA, OSHA, etc.) *and* making the workplace as safe as possible. However, while this group exists for their chartered purpose, very few who work in the facility believe their safety is totally in the hands of this safety department. It is well known in industry that *you* are the one most responsible for your own safety!

This is certainly the way healthcare workers would prefer it. However, the public believes their safety is the responsibility of individuals providing their healthcare. Oftentimes, people perceive it as rude, inappropriate, or condescending (especially in the very hospitable South) if they ask questions about what is going on with their care. One thing that should be understood is that patients should speak up and take control of their own care. After all, it is their lives that are at stake.

---

* Maxfield, David, Grenny, Joseph, McMillan, Ron, Patterson, Kerry, and Switzler, Al, *Silence Kills: The Seven Crucial Conversations for Healthcare,* VitalSmarts study, 7–8, 2005.

THE ROLE OF THE RISK MANAGER: INDUSTRY COMPARED TO HEALTHCARE

Understanding patient safety is everyone's responsibility; it must be accepted that individuals are obligated to be responsible in the organization for carrying out the programs that will assist in achieving patient safety goals and ensuring compliance with regulatory agencies. In healthcare, this role is primarily assumed by the title of *risk manager* (RM). While roles and responsibilities vary from hospital to hospital and system to system, the remainder of this topic will assume that the title of RM includes the responsibility for patient safety from the corporate standpoint. Such positions as *TJC compliance manager, patient safety officer, quality manager, performance improvement manager,* and *continuous improvement manager* are all titles that may share this role as well.

Having visited hospitals and attended conferences around the United States, it became apparent that a new vocabulary would have to be learned (as is the case with exploring any new working environment). Part of the frustrating task of this endeavor is not as much learning new words and acronyms as it is redefining current words in our respective internal dictionaries.

One of the key terms that had to be redefined was the title of *risk manager* (RM). Having met, talked with, and interviewed hundreds of RMs, a consistent description of roles and responsibilities could not be determined. The role and function of the healthcare RM depends upon the following: size of the organization (single hospital vs. large healthcare system), scope of services and activities, available resources, location of the organization to be served (rural vs. community), type of facility or organization (acute care, LTC), and the complexity of the organization (academic medical center).

The Office of Managing Risk and Public Safety (an office of the Department of the Interior)* defines risk management as:

> Simply put, Risk Management is a set of mitigation measures, including policies, and the associated decision making processes that reduce or eliminate risks. A risk is the likelihood of occurrence and severity of an adverse consequence of a hazard. Hazards are events, activities, and conditions that have the potential to cause harm.

However, as these wonderful people described their positions, we had to impose our definition of risk managers as compared to their roles in healthcare. Why is this significant?

In an effort to try to explain the differences between the role of the RM in healthcare and the role of the equivalent position in industry, some job postings for each were researched. The purpose was to compare the "average" job descriptions and explore the differences. It must be noted that the equivalent position of RM in healthcare to that of industry typically falls under the department of environment health and safety, or EH&S. Titles such as *quality manager, reliability manager,* and *maintenance manager* may also include such roles as risk.

---

* The Office of Managing Risk and Public Safety, What is Risk Management?

Here is the healthcare risk manager job posting found (all the references to a specific system have been removed):

MANAGER, RISK MANAGEMENT

The Risk Manager is responsible for the development, implementation, and management of the hospital's corporate risk management program which includes coordinating the placement of all insurance coverage, risk financing, managing claims against the facility, interfacing with defense legal counsel, medical staff, insurers, and brokers. The Risk Manager administers the risk management program on a day-to-day basis, serves as the Patient Safety Officer, managing and analyzing data, conducting educational programs, and complying with the governmental regulations and TJC standards. A Bachelor's degree and a minimum of five years experience in Risk Management, Safety, Insurance Management, Loss Prevention or related discipline required. Previous experience in clinical or hospital setting is preferred.

Here is the manufacturing equivalent of a similar job opening:

Certified Quality Engineer (Reagent Manufacturing)—5+ Years Experience, Degree, Pharmaceutical

Certified Quality Engineer—Reagent Manufacturing Job Duties
    25%—Implement cost of quality concepts, including quality cost categories, data collection, reporting etc. for reagents and reagent manufacturing.
    25%—Implement a systematic program for planning, controlling, and assuring product and process quality for reagents and reagent manufacturing.
    Processes for planning product and service development
    Material characterization
    Acceptance activities
    Measurement systems
    Product and process validation
    25%—Implement a systematic program for reliability and risk management for reagents and reagent manufacturing.
    Reliability life cycle characteristics
    Design for reliability
    Reliability and maintainability of reagents
    Reliability failure analysis and reporting
    25%—Implement a systematic program for quality problem solving and continuous improvement in reagents and reagent manufacturing.
    Quality improvement models such as Kaizen, plan-do-study-act, TQM, etc.
    Corrective and preventive action
    Barriers to quality improvement

Minimum Job Requirements: (Skills, Knowledge, Education, Experience, Certificates)

Minimum of five years experience in the pharmaceutical industry.

Minimum of five years experience as a certified quality engineer.

Bachelor's or master's degree in a life science or chemistry discipline or equivalent.

ASQ Quality Engineering Certified for at least five years.

Thorough understanding of quality philosophies, principles, systems, methods, tools, standards, organizational and team dynamics, customer expectations and satisfaction, supplier relations and performance, leadership, training, interpersonal relationships, improvement systems, and professional ethics.

Thorough understanding of a quality system and its development, documentation, and implementation with respect to domestic and international standards and requirements.

Thorough understanding of the audit process, including types of audits, planning, preparation, execution, reporting results, and follow-up.

Able to develop and implement quality programs, including tracking, analyzing, reporting, and problem solving.

Able to plan, control, and assure product and process quality in accordance with quality principles, which include software validation, product and process validation, planning processes, material control, acceptance sampling, and measurement systems.

Thorough knowledge of statistical analysis, reliability, maintainability, and risk management, including key terms and definitions, modeling, systems design, assessment tools, and reporting.

Thorough understanding of problem solving and quality improvement tools and techniques, including management and planning tools, preventive and corrective actions, and how to overcome barriers to quality improvements.

Able to acquire and analyze data using appropriate standard quantitative and statistical methods across a spectrum of business environments to facilitate process analysis and improvements.

Knowledge, Skills, and Abilities:

Specific knowledge of quality systems and design control for reagents and biological systems.

Excellent oral and written communication skills.

High degree of initiative with the ability to work independently.

Good use of process and business judgment.

Ability to plan and organize work while remaining flexible.

High degree of accuracy and quality.

Ability to work with all levels of employees.

Knowledge of word processing and spreadsheets.

While these two descriptions are not manufactured, they suggest an unfair comparison because of the contrast in level of detail. Much of this can be attributed to what people perceive about engineers and their tendency toward "paralysis by analysis."

What can be seen as the key differences in these two roles? In healthcare, in many instances the role of the risk manager is geared toward compensating for bad outcomes when they occur, or *risk compensation*. Hence, their responsibilities are geared toward ensuring that proper insurance levels are maintained, dealing with claims brought forth, and implementing systems to prevent recurrence. Another noted contrast between these two positions is that the RM, more often than the engineer, would place a lower weight on compensation in the decision to take a position. This clearly shows that they are not in it solely for money but for more humane reasons and aspirations to patient safety.

Contrast this to the manufacturing industry, where their risk management roles focus on identifying and quantifying the risks that exist in their environment (or *risk reduction*), and then using technologies available to minimize the risk. In short, their job is to make sure the incident does not occur, as opposed to compensating for them when they do occur. This is quite a divide and is quite in line with the healthcare culture.

This gets back to the basic paradigms associated with a culture. Responding to events only after they occur is a reactive approach that is typically very costly. Taking measures to eliminate or considerably reduce the potential of such events is a proactive approach and is much more economical in the big picture. This is a key fundamental principle of reliability, the distinction between proaction and reaction.

Over the past 30 years, a growing majority of manufacturers have adopted proactive approaches in their operations, which include the development and support of environmental health and safety (EH&S) departments, reliability departments, and quality departments. Within these departments, they have systems set up for preventive (time-based) approaches, predictive (condition-based) approaches, and proactive approaches. Proactive approaches include tools such as RCA, FMEA, risk assessment, Lean Six Sigma, reliability centered maintenance (RCM), and much more. Industry has proven beyond a shadow of a doubt that such proactive activities increase productivity, decrease the need for maintenance, decrease injuries, and improve the quality of the product.

In 1997, a benchmark survey, Top Quartile Paper Mills, claimed an average return on investment (ROI) of 19:1, while Chemical Processing came in at 16:1, and Steel production came in at 18:1 for their maintenance and reliability initiatives.*

In healthcare, there is still much skepticism as to the value of the business case for reliability or for proactive activities, which are geared toward patient safety and care. Building the business case for patient safety should not be too difficult a task when we consider what is at stake—patient lives!

ERROR RATES AND TECHNOLOGY

When reviewing the IOM report and the reported underestimation of deaths attributed to medical error, the question must be asked "What makes the healthcare field more prone to such high errors than any other industry?"

---

* Shultz, John, The Business Case for Reliability, available at http://reliabilityweb.com/art04/business_case.htm, 2004.

Again, reflecting on the history of industry over the past century, we found our-selves in the Industrial Age. This was a time when machines were evolving to help man accomplish work of various types. However, much of the work was still based on the brawn and physical stature of the man. The term "man" is used because, at the time, there were few women in the industrial workplace because of the physical requirements required. At this time, industry was very labor intensive and required much man-to-man interfacing. Error rates for industry during this time period could not be found, since the primary recorder of such information, the Occupational Safety and Health Administration (OSHA), was not created until 1971.

Moving more toward the information age, where the equipment became very sophisticated (e.g., robotics), it took less and less muscle and more and more brain-power to be multiple times more productive. This era saw the influx of women into the workplace.

Given this evolution, there is a heavy "person–machine" interface in industry today. As compared with the industrial revolution era, today in industry, three times more output is achieved with one third the manpower. As robotics becomes more sophisticated, the human will serve in a support role to the technology.

Now revert back to healthcare and curiosity about why error rates would be so high in such an industry. Again, another paradigm about healthcare from an outsider's perspective is that the technology used is of "Star Wars" caliber in healthcare. This has proven to be a half-truth. High technology in healthcare is employed on the sharp end or the medical diagnostics end, but *not* in the administrative end. Remember, this is using generalities, but this is what an industry outsider's observation has been.

The administrative infrastructure of a hospital today is reminiscent of that of industry 15 to 20 years ago. Experience shows that when the reliability approach* was introduced into industry 35+ years ago, there was the same resistance that can be seen today in healthcare. The information infrastructure in healthcare is archaic compared to the technologies that are available on the market today. Why?

Healthcare is a very labor-intensive industry where success depends and relies on responsible handoffs and good communication between caregivers, and yet they are expected to deliver care in an environment of archaic information systems.

Anytime there is a very labor-intensive industry that requires massive "human-to-human interfacing," the risk of human error is significantly increased. The risk of human error is reduced in industries where man primarily interfaces with machines, and reliability is highest when machines talk to machines. Remember, we are human, and with that come human foibles!

REGULATORY COMPLIANCE VERSUS PATIENT SAFETY—ARE THEY THE SAME?

In the previous section, it was established that error rates in healthcare are expected to be higher because of the dependency on "person–person" communications. This means that conditions are ripe for a breeding ground of human error.

---

* Latino, Charles J., *Strive for Excellence, The Reliability Approach,* Allied Chemical Corporation, 1980.

Knowing this, accreditation agencies such as TJC, and patient safety organizations such as the Institute for Healthcare Improvement (IHI), implement various guidelines to assist in ensuring that such risk of human error in decision-making is minimized. These are often well-intentioned acts that are designed to be in the best interest of patient safety. However, how can this good intent be misconstrued and abused to the point at which it is counterproductive to patient safety?

It is very common practice that when TJC either schedules an audit or arrives for a surprise audit, the tension in the facility quantumly elevates. Stress abounds, and all are trying to get their "ducks in a row." People scatter to get documentation in place, and everyone is put on notice of the visit to ensure that proper practices and housekeeping measures are in place and utilized. What is most important to notice here is that whatever priorities the facility was working on before the audit have now been dropped or placed on a back burner for the time being.

There is little doubt that the primary objective of such an audit, from the facility's perspective, is that they pass the audit with flying colors and maintain their accreditation status. This keeps the necessary federal funds flowing in order to keep the hospital running. From TJC's standpoint, auditors arrive with the greatest of intentions and a multitude of questions and checklists. If a particular auditor is satisfied that the facility has met or exceeded expectations, the recommendation will be that its accreditation status remain intact.

To complicate this matter even further, the General Accounting Office (GAO) released a report* outlining their findings regarding an audit done on TJC's performance. The highlights of this audit are summarized below:

> In a sample of 500 TJC-accredited hospitals, state agency validation surveys conducted in fiscal years 2000 through 2002 identified 31 percent (157 hospitals) with deficiencies in Medicare requirements. Of these 157 hospitals, TJC did not identify 78 percent (123 hospitals) as having deficiencies in Medicare requirements. For the same validation survey sample, TJC also did not identify the majority, about 69 percent, of deficiencies in Medicare requirements found by state agencies.

Now the auditing agency has been audited, and it has been found to be significantly deficient in identifying whether hospitals are in actual compliance with federal Medicare requirements or not.

What appears to have been lost in the translation is patient safety.

1. How does a successful audit directly correlate to actual patient safety numbers?
2. How is it actually measured that a patient is any safer in a facility that has passed an audit vs. one that has not?
3. Are such audits correlated to actual patient safety or more a "snapshot in time" that a facility passed an inspection?
4. Who is measuring actual patient safety—when a patient leaves in better condition than when arriving at the hospital?

---

* GAO Report, MEDICARE—CMS Needs Additional Authority to Adequately Oversee Patient Safety in Hospitals (GAO-04-850), Washington, DC, 2004.

5. Can a facility pass an audit without demonstrating an effective patient safety program?

The intent here is to demonstrate a disconnect between measurement and actual performance. Does this actually happen?

Here are some examples:

1. One facility's surveyor asks to review an RCA, while another facility's auditor does not.
2. A surveyor reviews an RCA for three minutes and determines it looks like all is order. However, it is not identified that all of the hypotheses in the analysis are validated with only hearsay rather than hard evidence.
3. One surveyor accepts an RCA using a fishbone diagram, while another does not for the same event.
4. A surveyor may request an RCA on a near miss rather than a sentinel event, while another auditor only expresses an interest in an RCA associated with a sentinel event.

The point here is that there can be a lack of continuity in how such surveys are done, as well as difference in the objectives. The GAO report confirms this. This is not saying this happens everywhere with every surveyor, but it does happen more often than it should.

When such disconnects occur, the patient loses out. This is because the patient may do well-intentioned research on a facility's record in the past and see that it is accredited. However, that accreditation may be misleading because the actual patient safety performance measurement is poor. How would the patient know? The patient is relying on the information from the regulatory agency to be accurate to make informed decisions. As will be discussed in this text, when efforts are made to measure the effectiveness of root cause analysis not by an institution's ability to be compliant but by its actual impact on patient safety, then reporting accreditation status will reflect a more accurate picture of a hospital's true performance in the arena of patient safety.

In a recent presentation made to a group of hospital CEOs, such concerns were raised. The sums of the responses were equivalent to, "We do not put much stock in TJC compliance. We feel if we focus on patient safety, then such compliance should be a by-product. If we have made our patients much safer and we are not in compliance, then there is something wrong with the compliance process."

Notice the difference in perspective from the CEO to the RM. The CEOs do not put much stock in the compliance efforts; however, the compliance effort consumes the RM at the sharp end.

This divide must be brought together with a common purpose and vision of absolute patient safety.

2 Creating Management Support for a Proactive Environment To Succeed

OBSTACLES TO LEARNING FROM THINGS THAT GO WRONG

In a recent informal online poll* presented to a group of beginner and veteran root cause analysis (RCA) practitioners, the following question was asked on the RCA discussion forum: What are the obstacles to learning from things that go wrong?

The following list is a summary of the responses, grouped into appropriate categories by the moderator. Some examples of the actual responses are listed below each category to help define what was meant by the category title.

1. RCA is almost contrary to human nature—28%.
   a. People don't like to admit they made the mistake.
   b. Accountability. If you are the boss—that is it!
   c. We are unwilling to change our own behavior.

2. Incentives and/or priority to do RCA's are lacking—19%.
   a. It is not expected of them.
   b. There is no personal incentive to do so.
   c. The work environment neither condones nor accommodates such a proactive activity.

3. RCA takes time/we have no time—14%.
   a. People are too busy due to daily work/problems.
   b. Variations on "I'm too busy."

4. Ill or misdefined RCA processes—12%.
   a. No agreement on "how far back" you have to go in your analysis.
   b. Vaguely defined processes.
   c. It is a theoretical approach. It is practically impossible.

5. Our "Western Culture"—9%.
   a. The stock market—short-term focus.
   b. Managers being rewarded for short-term results.
   c. The tyranny of the urgent.

---

* Nelms, Robert, What are the Obstacles To Learning From Things that Go Wrong?, available at http://www.rootcauselive.com, 2004.

6. We haven't had to do RCA in the past. Why now?—8%.
    a. Not how I was trained, not how I/we do things.
    b. Some behavior is so entrenched that it would be like being struck by lightning for some individuals to be aware of the need.

7. Most people don't understand how important it is to learn from things that go wrong—5%.
    a. It never occurs to most people that learning from experience is a cost-effective activity.

8. RCAs are not my responsibility—5%.
    a. It's NIMBY (not in my back yard).
    b. That's not our job.

The previous poll was cited to make an extremely important point to executives. As one can see from the list, every single objection is the result of an improper, inadequate, or nonexistent management support structure. Every one of these objections can be overcome with proper strategy, development, and implementation of a support structure.

Conversely, not addressing the leadership support structure will likely make such proactive efforts a lip-service exercise that is not capable of producing substantial results. An organization can have the best analysts and the best tools, but, without proper support from its leaders, the proactive efforts are not likely to succeed.

The following is a training model developed by Reliability Center, Inc. (RCI)* to provide guidance for the design and implementation of a support infrastructure for proactive activities such as RCA. It encompasses not only the elements about specific training objectives necessary to be successful, but it also outlines the specific requirements of the executives/management, the "champions," and the "drivers" who are accountable for creating the environment for RCA to be successful.

Specific information will be outlined from this model that is pertinent to creating the environment for RCA to succeed. For the sake of this text, we will focus on RCA being the primary proactive activity to support. However, the reader will recognize that the model will fit any proactive initiative.

THE ROLE OF EXECUTIVE MANAGEMENT IN SUPPORTING RCA

Like any new initiative trying to be implemented into an organization, the path of least resistance is typically from the top down, in contrast to the bottom-up approach. The one thing one should always be cognizant of is the fact that no matter what the new initiative is, it will likely be viewed by the end user as the "program of the month." This should always be in the back of our minds in developing implementation strategies.

---

* Reliability Center, Inc., *The Reliability Performance Process (TRPP)*, Reliability Center, Inc., Hopewell, VA, 2004.

Experience demonstrates that the closer we get to the sharp end, where the work is actually performed, the more skeptics will be encountered. Every year, a new organizational "buzzword" fad emerges, and the executives hear and read about them in trade journals, magazines, business text, and from colleagues. Eventually, directives are given to implement these "fads," and by the time it reaches the sharp end, the well-intentioned objectives of the initiatives are so diluted and saturated from miscommunication they are viewed as non-value-added (NVA) work and an added burden to an existing workload. This is the paradigm of the end user that must be overcome to be successful at implementing RCA. The recent adaptation of Six Sigma to healthcare is an example of this scenario and will be expanded on in later chapters.

Oftentimes, when looking at instituting these types of initiatives, they are looked at strictly from the shareholders' view and work backward. Do not get us wrong—we are not against new initiatives that are designed to change behavior for the betterment of the corporation. This process is necessary to progress as a society. However, the manner in which the organization tries to attain the end is what has been typically ineffective.

The organization must look at linking what is different about this initiative, from the perception of the field or end user, as opposed to others that have been tried and unable to succeed. The reality of the environment of the people who will make the change happen must be considered. How can the behavior of a given population be changed to reflect behaviors that are necessary to meet the organization's objectives?

For example, consider a maintenance person in a hospital's facility engineering department. This person is expected to repair equipment so that the hospital can run more efficiently, allowing more patients to be seen, thus increasing profits. As a matter of fact, performance is measured by how well the individual can make the repair in the shortest time frame possible. Such employees are given recognition when emergencies occur and they respond almost heroically.

Now comes along this RCA initiative, and the organization wants maintenance workers to participate in making sure failures do not occur anymore. In their mind, if this objective is accomplished, they are out of their jobs! Rather than be perceived as *not* being team players, they will superficially participate until the "program of the month" has lived out its normal six-month life and then go on with business as usual. This scenario occurs repeatedly, and it is a very valid concern based on the reality of the end user. This perception must be overcome prior to implementing an RCA initiative.

Maintenance, in its true state, is viewed as a necessary evil to an organization. Consequently, when equipment fails, it generally holds up facility operations, which affects patient safety and quality, which holds up profitability. Imagine a world where the only failures that occurred were predictable. This is a world that organizations must move toward as precision environments become more the expectation. As efforts move in this direction, there will be less need for maintenance-type skills on a routine basis.

What about the area of reliability? Most organizations never seem to have the resources to properly staff their reliability, safety, quality, risk, and/or performance improvement groups. There are plenty of available roles in these proactive disciplines. Think about how many proactive jobs would be available if there were money and people able to fill them: FMEA analysts, RCA analysts, Six Sigma black belts, TPM analysts, risk analysts, data trenders, inspectors/monitors, patient safety officers, etc.

As trainers, we are continually intrigued by the most frequently used objection to using RCA from our students: "I don't have time to do RCA." If one thinks hard about this statement, it really is an oxymoron. Why do people typically *not* have time to do RCA? They are so busy fire fighting (being reactive) that they do not have time to analyze why the event occurred in the first place. If this remains as a management strategy, then the organization will never progress, because no level of dedication is being put toward "getting rid of the need to attend to the reactive work!"

So how can executives get these very same people to willingly participate in a new RCA initiative (Figure 2.1)?

1. It must start with an executive putting a rubber stamp on the RCA effort, outlining specifically what the expectations are for the process and setting a time line for seeing bottom-line results.
2. The approving executives should be educated in the RCA process themselves, even if it is an overview version. Such demonstrations of support are worth their weight in gold, because the users can be assured that the executives have learned what they are learning and agree with and support the process.
3. The executive responsible for the success of the effort should designate a "champion" of the RCA effort. This individual's roles will be outlined later in this chapter.
4. It should be clearly delineated how this RCA will benefit the company, but more importantly it should also delineate how it will benefit the work life of every employee.

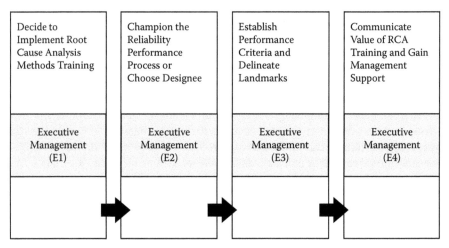

FIGURE 2.1 Executive management support roles.

5. Next the executive should outline how the RCA process will be implemented to accomplish the objectives and how management will support those actions.

6. A policy and/or procedure should be developed to institutionalize the RCA process. This is another physical demonstration of support that also provides continuity of the RCA application and perceived staying power. It gives perceived staying power to the effort because, even if there is a turnover in management, institutionalized processes have a greater chance of weathering the storm. A policy and procedure (P&P) will provide the drivers the clout they need when conducting RCAs. When physicians do not want to participate on RCA teams and management wants to change the findings because they do not like what was uncovered, the driver has to be able to respond that any lack of cooperation with RCAs is a violation of company policy.

7. However, the most important action an executive can take to demonstrate support is to sign a check. This is a universal sign of support.

THE ROLE OF AN RCA CHAMPION

Even if all of the above actions take place, this does not automatically ensure success. How many times have well-intentioned efforts from the top tried to make their way to the sharp end and failed miserably? Typically, somewhere in the middle of the organization, the translation of the original message begins to deviate from its intended path. The miscommunication of the original message is a common reason why some very good efforts fail!

If organization members are proactive in their thinking, and they foresee such a barrier to success, then they can plan for its occurrence and avoid it. This is where the role of the RCA champion comes into play (Figure 2.2).

There are three major roles of an RCA champion:

1. The champion must administer and support the RCA effort from a management standpoint. This includes ensuring that the message is communicated properly and effectively from the top to the floor. Any deviations from the plan will be the responsibility of the champion to align or get back on track. This person is truly the "champion" of the RCA effort.

2. The second primary role of the RCA champion is to be a mentor to the drivers and the analysts. This means that the champion must be educated in the RCA process and have a thorough understanding of what is necessary for success.

3. The third primary role of the RCA champion is to be a protector of those who utilize the process and uncover causes that may be politically sensitive. Sometimes this role is referred to as providing "air cover" for ground troops. In order to fulfill this responsibility, the RCA champion must be in a position of authority in order to take a defense position and protect the person who uncovered these facts.

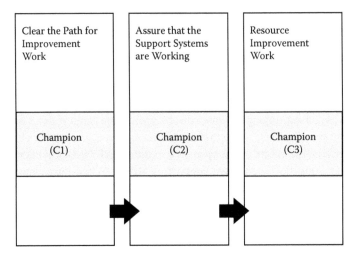

FIGURE 2.2 RCA champion support roles.

Ideally this would be a full-time position. However, in reality, typically this is a part-time effort for an individual. The key is that it must be made a priority to the organization. This is generally accomplished if the executive(s) perform their designed tasks set out above. Actions do speak louder than words. When new initiatives come down the pike and the workforce sees no support, then it becomes another "they are not going to walk-the-talk" issue. These are viewed as lip-service programs that will pass over time. If the RCA effort is going to succeed, it must first break down these types of paradigms that currently exist. It must be viewed more different than the other programs. This is also the RCA champion's role in projecting an image that this is different and will work.

The RCA champion's additional responsibilities include ensuring that the following responsibilities are carried out (Figure 2.3):

1. Selecting and training RCA drivers who will lead RCA teams. What personal characteristics and traits are required to make this a success? What kind of training do they need to provide them the tools to do the job right?
2. Developing management support systems such as:

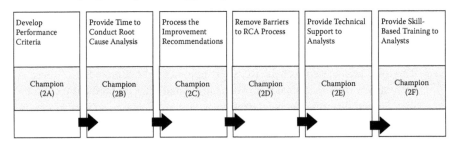

FIGURE 2.3 Additional roles of RCA champions/management.

a. *RCA performance criteria*—What are the expectations of financial returns that are expected from the organization? What are the time frames? What are the landmarks? How do the results correlate to patient safety?

b. *Providing time*—In an era of scarce resources and lean staffs, "How are we going to mandate that designated employees *will* spend XX% of their week on RCA teams?"

c. *Process the recommendations*—How are recommendations from RCAs going to be handled in the current reactive budgetary system? How does improvement (proactive) work get executed in a reactive budgetary system?

d. *Provide technical/clinical resources*—What technical/clinical resources are going to be made available to the analysts to help them prove and/or disprove their hypotheses using the "whatever it takes" mentality?

e. *Provide skill-based training*—How will RCA team members be educated and how to ensure that they are competent to participate on such a team?

3. The champion shall also be responsible for setting performance expectations. The champion should draft a letter that will be forwarded to all students who attend the RCA training. The letter should clearly outline exactly what is expected of them and how the follow-up system will be implemented.

4. The champion should ensure that all training classes are kicked off either personally, by an executive, or by another person of authority, giving credibility and priority to the effort.

5. The champion should also be responsible for developing and setting up a recognition system for RCA successes. Recognition can be in the form of a letter from an executive to gift certificates for nice dinners. Whatever the incentive is, it should be of value to the recipient.

Needless to say, the role of a champion is very critical to the RCA process. The lack of a champion is usually why most formal RCA efforts fail. In this case, there is no one leading the cause or carrying the RCA flag. Make no bones about it, if an organization has never had a formal RCA effort, or had one and failed, such an endeavor is an uphill battle. Without an RCA champion, sometimes it can seem like the analysts are on an island by themselves.

THE ROLE OF THE RCA DRIVER

The RCA driver can be synonymous with the RCA team leaders. These are the people who organize all the details and are closest to the work. Drivers carry the burden of producing bottom-line results for the RCA effort. Their teams will meet, analyze, hypothesize, verify, and draw factual conclusions as to why undesirable outcomes occur. Then they will develop recommendations or countermeasures to eliminate the risk of recurrence of the event (Figure 2.4).

All the executives', managers', and champions' efforts to support RCA are directed at supporting the driver's role to ensure success. Drivers are in a unique

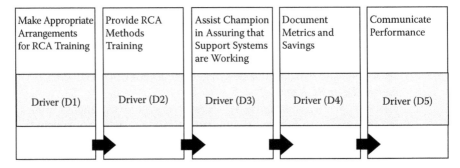

FIGURE 2.4   RCA driver support roles.

position in that they deal directly with the experts—the people who will compose the core team. The personality traits that are most effective in this role, as well as a core team member role, will be discussed at length in Chapter 8, Ordering the Analysis Team.

From a functional standpoint the RCA drivers' roles are:

1. *Making arrangements for RCA training for team leaders and team members*—This includes setting up meeting times, approving training objectives, and providing adequate training rooms.
2. *Reiterating expectations to students*—Clarify to students what is expected of them, when it is expected, and how it will be obtained. The driver should occasionally set and hold RCA class reunions. This reunion should be announced at the initial training so as to set an expectation of demonstrable performance by that time.
3. *Ensure that RCA support systems are working*—Notify the RCA champion of any deficiencies in support systems and see that they are corrected.
4. *Facilitate RCA teams*—The driver shall lead the RCA teams and be responsible and accountable for the team's performance. The driver will be responsible for properly documenting every phase of the analysis.
5. *Document performance*—The driver will be responsible for developing the appropriate metrics to measure performance against. This performance shall always be converted from units to dollars when demonstrating savings, hence success. Success should also be documented based on a direct correlation to an improvement in patient safety. Efforts should be taken to also correlate improvements in patient safety to bottom-line financial performance.
6. *Communicate performance*—The driver shall be the chief spokesperson for the team. The driver will present management updates as well as other individuals on site and at other similar operations that could benefit from the information. The driver shall develop proper information distribution routes so that the RCA results get to others in the organization who may have, or have had, similar occurrences.

The driver is the last of the support mechanisms that should be in place to support such an RCA effort. Most RCA efforts we have encountered are put together at the last minute as a result of an "incident" that just occurred. This topic was discussed earlier regarding using RCA as only a reactive tool.

A structured RCA effort should be properly placed in an organizational chart, because RCA is intended to be a proactive task. It should reside under the control of a structured proactive department such as risk management, patient safety, reliability, quality, or the like. In the absence of such a proactive department, it should report to a staff position such as a VP of operations or VP of quality.

Whatever the case may be, ensure that an RCA effort is never placed under the control of a reactive department. By its nature, a reactive entity's role is to respond to the day-to-day activities and problems in the work environment. For instance, in industry, the role of a true reliability department is to look at tomorrow, not today. Any proactive task assigned to a reactive department is typically doomed from the start.

This is the reason that, when "reliability" became the industry buzzword of the mid 1990s, many industrial maintenance engineering departments were renamed as reliability departments. The same people resided in the department, and they were performing the same jobs, however their title was changed and not their function (Figure 2.5).

If you are an individual who is charged with the responsibility of responding to daily problems and also seizing future opportunities, you are likely to never get to realize those opportunities. Reaction wins every time in this scenario.

Now assume that at this point the organization has developed all the necessary systems and personnel to support an effective RCA effort. How do they know what opportunities to work on first? Working on the wrong events can be counterproductive and yield poor results. In the coming chapters on failure mode and effects analysis (FMEA) and opportunity analysis (OA), techniques will be discussed to help sell why the analyst should work on one event versus another.

SETTING FINANCIAL EXPECTATIONS: THE REALITY OF THE RETURN

As discussed earlier, one of the roles of the champion is to delineate financial expectations of the RCA effort. This will obviously vary with the key performance indicators (KPIs) of each organization, but in this section we will look at providing a typical business case to justify implementing an RCA effort.

Because the costs to implement such an effort will vary based on each facility, their product sales margin, their labor costs, and the training costs (in-house versus contract), we will base our justifications on the following assumptions:

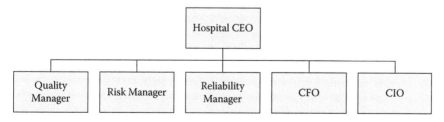

FIGURE 2.5   Ideal position for reliability or equivalent proactive department.

1. Assumptions
   a.   Loaded cost of hourly employee—$US 50,000/yr.
   b.   Hourly employees will spend 10% of their time on RCA teams.
   c.   Loaded cost of full-time RCA driver (salaried) —$US 70,000/yr.
   d.   RCA driver will be a full-time position.
   e.   RCA training costs (hourly) —$US 400/person/day.
   f.   RCA training costs (salaried) —$US 500/person/day.
   g.   Population trained—per 100 trained.

2. RCA Return Expectations
   a.   Train 100 hourly employees in RCA methods. (Keep in mind that, while most of these individuals will not lead RCA teams, they will participate on them from time to time. Also, good RCA training will teach the students about proactive human error reduction techniques and how to prevent errors by recognizing conditions where errors are being initiated.)
   b.   Train one salaried employee to lead RCA effort.
   c.   Critical Mass (assumption): 30% of those trained will actually use the RCA method in the field. This results in 30 personnel trained in RCA methods actually applying in the field (100 trained × 30% applying).
   d.   Of the 30 personnel applying the RCA method, let's assume they are working in teams of three at a minimum. This results in 10 RCA teams applying the methodology in the field (30 personnel/3 per team).
   e.   Each RCA team will complete one analysis every two months. This results in 60 completed analyses per year (10 RCA teams × 6 analyses/yr)
   f.   Each "significant few" (to be discussed in opportunity analysis chapter) analysis will net a minimum of $US 50,000 annually. This results in an annual return of $US 3,000,000 per 100 people trained in RCA methods.

3. The Costs of Implementing RCA
      YEAR 1
      a.   Training 100 hourly employees in 3 days of RCA—$US 120,000.
      b.   Training 1 salaried person in 5 days of RCA—$US 2,500.
      c.   10% of 30 hourly employees time per week, annually—$US 150,000.
      d.   Salary of RCA driver/year—$US 70,000.
      e.   Total RCA implementation costs for year 1—$US 342,500.

      YEAR 2
      a.   Training 100 hourly employees in 3 days of RCA—$US 0.
      b.   Training 1 salaried person in 5 days of RCA—$US 0.
      c.   10% of 30 hourly employees time per week, annually—$US 150,000.
      d.   Salary of RCA driver/year—$US 70,000.
      e.   Total RCA implementation costs for year 1—$US 220,000.*

* All costs of resources to prove hypotheses and implement recommendations are considered as sunk costs. Technical resources are currently available and budgeted for, regardless of RCA. Also, recommendations from RCA generally result in the implementation of organizational system corrections; for instance, rewriting procedures, providing training, upgrading testing tools, restructuring incentives, etc. These types of recommendations are not generally considered as capital costs. Capital costs resulting from RCA, in our experience, are not the norm but the exception.

4. Return on Investment
   YEAR 1
      a. Total expected return, year 1—$US 1,500,000.*
      b. Total expected costs, year 1—$US 342,500.
      c. ROI, year 1—437%.

* Assumes that it will take six months to train all involved and get up to speed with actually implementing RCA and the associated recommendations. This is the reasoning for cutting this expectation in half for the first year.

   YEAR 2
      a. Total expected return, year 2—$US 3,000,000
      b. Total expected costs, year 2—$US 220,000
      c. ROI, year 2—1360%

As we can tell from these numbers, the opportunities are left to the imagination. They are real. They are phenomenal to the point that they are unbelievable. When the process just discussed is gone through, look at the conservativeness built in:

1. Only 30% of those trained will actually apply the RCA method.
2. Students will only spend 10% of their time on RCA.
3. Students will work in teams of three or more.
4. Students will only complete one RCA every two months.
5. Each event will only net $US 50,000/year.

Use this same cost–benefit thought process and plug in your own numbers to see if the ROIs are any less impressive. Using the most conservative stance, it would appear irrational *not* to perform RCA in the field. How many corporate capital projects would be turned down if they were demonstrated to management to have an ROI ranging from 437% to 1360%? Not many!

II

Event Prioritization Techniques

3 Failure Classification

PROBLEMS VERSUS OPPORTUNITIES

To begin discussing the issue of root cause analysis (RCA), the foundation must be laid with some common understanding of key terminology. For instance, what are the key differences between problems and opportunities? Many people tend to use these terms interchangeably. The truth is, these terms are really at opposite ends of the spectrum in their definitions.

A problem can be defined as *a deviation from some performance norm* (see Figure 3.1). What exactly does this mean? It simply means the normal level or standard an organization is used to cannot be met. For example, look at an emergency department (ED) in a moderate-sized hospital. One of the goals of the ED is to ensure that if a patient arrives with symptoms of a heart attack, he is given aspirin immediately. The average rate for providing potential heart attack patients an aspirin immediately upon arrival could be 85% of the time. It would be a problem if the ED lost power and, in the ensuing chaos, those who should have gotten aspirin did not. This may drop the average rate to, say, 80%, which is totally unacceptable.

An opportunity is really just the opposite of a problem (see Figure 3.2). It can be defined as *a chance to achieve a goal or an ideal state.* This means that some changes are going to be made to increase the performance norm above the status quo. Continue with the aspirin issue described above. If such potential heart attack patients were getting an aspirin immediately upon arrival 85% of the time, then an opportunity would be to raise that rate to 100% of the time. In this case, the reasons that such patients were not getting the aspirin would be analyzed, and recommendations would be implemented to ensure that all such patients got the aspirin at arrival. Attaining an average rate above the status quo would be deemed an opportunity.

Now these terms need to be put into perspective. When a problem occurs and action is taken to fix it, does the system actually improve or progress? The answer to this question is an emphatic *no.* When working on problems, an organization is essentially working to maintain the status quo or performance norm. This is synonymous with the term "reaction." When a problem occurs, the organization reacts to get things back to their normal state. If all of our efforts are directed at working on problems, the organization will never be able to progress. To prove this, just ask yourself how much time you spend reacting versus proacting in your job responsibilities. The general consensus is that about 80% of our time is spent reacting and 20% is spent proacting. If this is true, then there is very little progress being made. This would seem to be

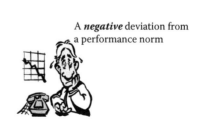

A *negative* deviation from a performance norm

FIGURE 3.1   Problem definition graph.

A chance to achieve a goal or an ideal state.

FIGURE 3.2   Opportunity definition graph.

a key indicator as to why most productivity increases are minimal from year to year.

Consider opportunities for a moment. When working on opportunities does the organization concentrate efforts in this area? The answer is *yes*! When opportunities are achieved, the organization will be striving to raise the status quo to a higher level. Therefore, to progress, an organization would have to begin taking advantage of the numerous opportunities presented to it. So if working on problems is like reacting, then working on opportunities is like proacting (see Figure 3.3).

So the answer is simple. The organization should start working on opportunities and disregard problems, right? Why can't this be done? There are many reasons, but a few are obvious. Problems are more obvious since they take away from normal operations. Therefore, more attention and priority are given to them. Opportunities can always be put off until tomorrow, but problems have to be addressed today. There is also the issue of rewards. People who are good reactors, who come in and save the day, tend to get pats on the back and the old "atta-boys." What a great thing from the reactor's perspective—recognition, overtime pay, and most importantly, job security. Often the person who tries to prevent a problem or event from occurring gets the cold shoulder, while the person who comes in after the event has occurred gets treated like royalty. This is not to say that good reactors should not be rewarded for duties above and beyond their call, but proactive behavior should be rewarded as well.

Then there is the risk factor. Which are more risky, problems or opportunities? Opportunities are always more risky, since there are many unknowns. With problems, there are virtually no unknowns. The problems have likely been fixed before, so there certainly is the confidence to fix it again. An old saying that comes to mind here is, "…when you get really good at *fixing* something, you should wonder why you are getting so much practice."

In a perfect world, one should have to pull the manual out to see what steps to take to fix the problem. How many times does one see a nurse or physician pulling out the manual to troubleshoot a problem? People today do not want to take a lot of chances with their careers, so opportunities begin to look like what are often called "career-limiting" activities.

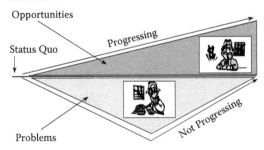

FIGURE 3.3   The difference.

So with that said, changing the paradigm that reactive is always more important than proactive work must be achieved. This means opportunities are just as important as, if not more important than, problems.

CHRONIC VERSUS SPORADIC EVENTS

What are the basic different types of failures or events that can occur? There are two basic categories of failures that can exist: sporadic and chronic. Let's look at each of these categories in greater detail.

For the purposes of vocabulary and common understanding, the term *sporadic* shall follow J. M. Juran's definition:*

> A sporadic problem is a sudden adverse change in the status quo, requiring remedy through restoring the status quo.

A sporadic occurrence usually indicates that a dramatic event has occurred. For example, maybe the patient was overmedicated or a wrong-site surgery was performed. These types of events tend to demand a lot of attention—not just attention, but urgent and immediate attention. In other words, everyone in the organization knows that something bad has happened. The key characteristic of sporadic events is they rarely happen (hopefully!). There is one mechanism at work that has caused this event to occur. This is important to remember. Sporadic events have a dramatic impact when they occur, which is why financial figures tend to be applied to them. For instance, one may hear someone say, "We paid out $10,000,000 for that baby swap incident last year."

One of the significant initiatives in healthcare today is the drive to create a common, standardized, and accepted taxonomy of terms. Such efforts are all about how we capture events and categorize them. Recently, the National Quality Forum (NQF) endorsed The Joint Commission (TJC) taxonomy as a move toward this unified purpose. There are also other taxonomies, such as the National Coordinating Council for Medication Error Reporting and Prevention (NCC MERP) that many organizations have expanded beyond medication events. Under such taxonomies, the term "sporadic" events would be consistent with the healthcare terms of "never events" and "sentinel events."

Sporadic events are very important and they certainly do cost a great deal of money when they occur. The reality, however, is that they do not happen very often. If an organization has a lot of sporadic events then they certainly would not be able to survive from the business end, and their reputation in the community would be tarnished.

Again, for the purposes of vocabulary and common understanding, the term "chronic" shall also follow Juran's definition:

> A chronic problem is a long-standing adverse situation, requiring remedy through changing the status quo.

* Juran, M., and Gyrna, F. M., *Quality Planning and Analysis,* McGraw-Hill, New York, 1980, 99.

Chronic events, on the other hand, are not very dramatic when they occur. Such events happen so often that they become a cost of doing business. Organizations become so proficient at working on these events that they become part of the status quo. "Normal" output can be obtained in spite of these types of events.

What are the characteristics of chronic events? Chronic events are accepted as part of the routine. They are accepted as fact and they are going to happen. In our organization, such events will be accounted for in various department budgets. Such budgets are in place to make sure that, when routine events occur, there is money on hand to address them. These types of events do, however, demand attention, but usually not the attention a big sporadic event would require. The key characteristic of a chronic event is the frequency factor. These events happen over and over again for the same reason or mode. For instance, a given printer in a nursing station may fail three or four times a week, holding up vital patient-related information. This would be considered a chronic event. Chronic events tend not to get the same attention as sporadic events because they are usually not very costly individually, but their combined consequences can be very costly. Therefore, a dollar figure is rarely assigned to an individual chronic event.

Most people fail to realize the tremendous effect the frequency factor has on the cost of chronic failures. Take for instance blood redraws in an ED. When a redraw is required, it will take additional time of the nurse, lab tech, physician, and patient. Additional supplies such as syringes, gauze, band-aids, needles, etc. are required. Costs associated with using the real estate in the ED are also incurred.

An individual redraw may cause delays that average 60 minutes. That, in and of itself, may seem inconsequential. However, when the big picture is considered and it is determined that there are over 10,000 blood redraws per year in the ED, the event seems a little more important. Now, instead of a single 60-minute delay, we are looking at 10,000 occurrences that cost on average about $300 each. Now that little problem becomes a $3,000,000 per year problem.

This clearly demonstrates that the frequency factor is very powerful. The normal tendency is to only see chronic events in their individual state and to sometimes overlook the accumulated cost over a period of time. If one were to go into a facility and aggregate all of the chronic events over a year's time and multiply their effects by the number of occurrences the yearly losses would be staggering (and accepted as a cost of doing business).

How do chronic and sporadic events relate to the discussion on problems and opportunities? Sporadic events take the organization below the status quo and tend to take an extended period of time to restore. When the operation is restored, it simply returns to the status quo. This is very much like what happens when reacting to a problem. The problem occurs, and action is taken to get back to the status quo.

Chronic events, on the other hand, happen so routinely that they actually become part of the status quo. Therefore, when they occur, they do not take us below our performance norm. If, in turn, the chronic or repetitive events were to be eliminated, then the elimination would actually cause the status quo to improve. This improvement is the equivalent of realizing an opportunity. So by focusing on chronic events and eliminating the causes and not simply fixing the symptoms, the organization is

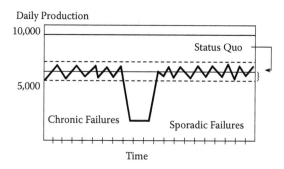

FIGURE 3.4   The linkage.

really working on opportunities. As discussed before, working on opportunities will progress the organization (see Figure 3.4).

Because chronic events can advance the organization, the significance of this must be discussed further. Sporadic events by their very nature are high-profile and high-cost events. But the costs of sporadic events can be amortized over a long period of time so that the effect is not as severe from an accounting standpoint.

Consider a claim against the hospital in which the insurance company opts to settle instead of taking it to verdict. Even under these conditions, there could be a portion of the settlement for which the hospital will be responsible (depending on insurance limits) and/or insurance premium increases as a result of the claims paid. Whatever the payout, it would likely be amortized over a longer period of time (like 5, 10, or 20 years), depending on the size of the payout.

Chronic events, on the other hand, have a relatively low impact on an individual basis, but their overall impact is often overlooked. If we were to aggregate all of the chronic events from a particular facility and look at their total cost over a one-year period, we would see that their impact is far more significant than any given sporadic event, simply due to the frequency factor.

As is well known, organizations are in business to make a profit—even nonprofit organizations. While nonprofit organizations cannot retain earnings, they do depend on profits to assist in adding value to their services, covering costs associated with patients who cannot pay, increasing patient safety efforts, and expanding their physical facilities.

When a sporadic event occurs, it affects the profitability of a facility significantly the year in which it occurs, but once the problem has been resolved, profitability gets back to "normal." The dilemma with chronic events is that they usually never get resolved, so they affect profitability year after year. If chronic events were eliminated (or greatly reduced) instead of just reacting to their symptoms, the organization would make great strides in profitability. Imagine if a healthcare system had 10 facilities and they were able to increase their revenues by 10% each. In essence, they would have the capacity equivalent to one new facility without spending the capital dollars to build a new hospital. That is the power of resolving chronic issues.

To sum up this discussion on failure classification, the key ideas presented will be reviewed. The world is full of problems and opportunities. We would all love to take advantage of every opportunity that comes about, but it seems as if there are

too many problems confronting us to take advantage of the opportunities. By eliminating the chronic failures, we are really achieving opportunities as well as adding
additional time to eliminate more problems. In the next chapter, we will discuss a
method for uncovering all of the events for a given process and delineating which of
those events are the most significant from a business perspective.

4 Basic Failure Mode and Effects Analysis
The Traditional Approach

This technique was first applied in the aerospace industry to determine what failure events *could* occur within a given system (e.g., a new aircraft) and what the associated effects would be if it were to occur. This technique, albeit effective, is very man-hour intensive. It is estimated that a typical failure mode and effects analysis (FMEA) in the aerospace industry takes anywhere from 50 to 100 man-years to perform on a new aircraft design. There are many good reasons that this technique takes so long to complete, and there are significant benefits to performing one in healthcare.

TJC Standard LD.5.2 requires facilities to select at least one high-risk process for proactive risk assessment each year.* This selection is to be based in part on information published periodically by TJC that identifies the most frequently occurring types of sentinel events.

The Department of Veterans Affairs (VA) National Center for Patient Safety (NCPS) also issued guidelines for conducting what they refer to as healthcare FMEAs or HFMEAs™.† One of the primary differences between these two FMEA approaches is that the HFMEA incorporates the use of a parameter referred to as "detectability," whereas the traditional use of FMEA does not require this parameter. The intent of this text is not to change or redefine the scope and purpose of the current FMEA requirements from a regulatory standpoint, but to bring to the attention of the analysts that these are "guidelines" and not requirements.

Regulatory bodies typically do not outline specific approaches or methodologies in their requirements because of potential conflict-of-interest concerns. If they were to do so, and it was to the detriment of one methodology over another, they would receive complaints about bias and discrimination. Therefore most guidelines and standards put the emphasis of their measurements on outcomes of the analysis (ends) as opposed to the approaches (means) to attain them.

Both TJC and the NCPS have proposed outlines for FMEA approaches that will, when properly used, comply with the regulatory requirements. However, if variations are used in these approaches and equivalent outcomes are attained, they should also comply with the guidelines and pass an audit.

Let's take a look at a brief overview of the minimum elements of a basic FMEA. The following approach is an aggregation of elements of FMEA from industry, from the TJC FMEA standards and from the NCPS HFMEA:

---

* TJC, *Sentinel Event Policy and Procedures*, 2004, available at http://www.jointcommission.org/ SentinelEvents/PolicyandProcedures.
† HFMEA is a pending trademark of the VA National Center for Patient Safety.

1. Identify the scope of the system.
2. Define an unacceptable risk in the system.
3. Organize the analysis team.
4. Establish severity ratings.
5. Establish probability ratings.
6. Establish detectability ratings (optional).
7. Complete FMEA spreadsheet.
   a. Define subsystem.
   b. Define event.
   c. Define mode.
8. Develop corrective action plan.

STEP 1: IDENTIFY THE SCOPE OF THE SYSTEM

The first step in the FMEA process is to identify the scope of the system to be analyzed and draw a basic process flow diagram (block diagram) of the subsystems within that system (see Figure 4.1). At this phase, a definition of loss will also have to be developed to focus the team on what they are looking for in terms of weaknesses in the system. In this case, the system chosen is a proposed ultrasound review process in an OB/GYN unit.

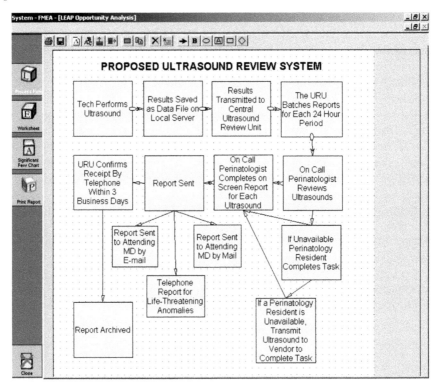

FIGURE 4.1 Proposed ultrasound review process—block diagram.

STEP 2: DEFINE AN UNACCEPTABLE RISK IN THE SYSTEM

Once a system has been chosen to analyze for unacceptable risks, it must be understood by the team what a "risk" is in that system. Therefore, we must define what is an unacceptable risk in the system chosen to analyze. When considering what a risk is, we are looking for something that could occur in our system that would result in an undesirable outcome. These risks usually manifest themselves in certain types failure modes. For instance, in our proposed ultrasound review process example, our definition of risk (i.e., loss) may be any event or condition resulting in an unacceptable delay that increases the risk of harm to the patient.

If we do not define what an unacceptable risk is in our system, then we leave the definition up to individual interpretation by each team member. Therefore, members from quality, risk, performance improvement, etc. will each be using their own definitions, which will eventually dilute the focus of the analysis and produce disjointed results.

STEP 3: ORGANIZE THE ANALYSIS TEAM

Next in the process, an appropriate team should be assigned to conduct the analysis. A principal analyst (PA) who will be accountable for, and facilitate, the FMEA process with the team members should be chosen.

Who are the appropriate team members for an FMEA? To answer this question, we should ask ourselves another question, "Who would be able to identify the specific risks in the given system?" Often, conducting such analyses is viewed as a managerial task (as it is often viewed in manufacturing as an engineering task). However, managers are most likely removed from the sharp end of the system and not as familiar with those who are in daily contact with the patient and/or system. It is imperative that sharp-end personnel be included on such teams, as they bring reality to the table. They bring experience with having faced such risks in the past and are invaluable to the credibility of such an analysis.

STEP 4: ESTABLISH SEVERITY RATINGS

The team must now establish an acceptable rating system for weighting the potential severity of the events that could occur with the system chosen. Figure 4.2 is a good sample of the one used by the NCPS.*

Since we are doing a probability analysis, we are trying to reasonably forecast what risks may materialize and then be able to prioritize those risks quantifiably. Conducting an FMEA can be somewhat frustrating, because we are dealing with events that *could* happen as opposed to those that *have* happened. Trying to quantify something that has not happened puts our trust in subjectivity. This is why these rating systems are so important, as they are the criteria in which our team members will base their estimates. The more reflective and accurate the rating scale, the more accurate the risk forecast.

---

* NCPS, *The Basics of Healthcare FMEA (HFMEA)*, 2004, available at http://www.va.gov/ncps/safety-topics/HFMEAIntro.pdf.

Catastrophic Event	10	Failure could cause death or injury
Major Event	7	Failure causes a high degree of customer dissatisfaction
Moderate Event	4	Failure can be overcome with modifications to the process or product, but there is minor performance loss
Minor Event	1	Failure would not be noticeable to the customer and would not affect delivery of the service or product

FIGURE 4.2 Sample severity rating scale—NCPS.

STEP 5: ESTABLISH PROBABILITY RATINGS

Now the team must establish an acceptable rating system for weighting the probability or likelihood of the events occurring within the system chosen. Figure 4.3 is a good sample of the one used by the NCPS.*

Because we are trying to project how often something *might* happen, we again have to put our trust in our rating scale and its appropriateness for the system we have chosen. While the NCPS scale may seem appropriate for some systems in which we apply the FMEA, it may serve merely as a benchmark document for others. For instance, if we are looking at a system, such as blood drawing, where frequencies of a certain type of event may be hourly or daily, our rating scales may need to be changed to reflect the chronic nature of these events.

STEP 6: ESTABLISH DETECTABILITY RATINGS

The use of "detectability" is advocated by the NCPS in their HFMEA approach. However, it is not required by TJC. Based on the experience of working with healthcare analysts, the field is split on the use of this parameter. The use of detectability seeks to evaluate the likelihood of potential occurrences of events identified in the FMEA to the current system.

The apparent advantage of using detectability is that it provides a safety net or another defense mechanism preventing the harm from reaching the patient. Knowing that this safety net is in place provides a sense of security and allows the organization time to identify and correct the situation.

The potential disadvantages are the same as the advantages in the sense that this could create a false sense of security. Opponents would argue that including detectability would hide the true risk of the potential event (severity × probability). After uncovering the raw risk of an event, if it is high, it can be addressed by either reducing or eliminating the chances of its occurrence instead of implementing additional systems to detect its consequences earlier.

---

* NCPS, *The Basics of Healthcare FMEA (HFMEA)*, 2004, available at http//www.va.gov/ncps/safety-topics/HFMEAIntro.pdf.

Frequent	4	Likely to occur immediately or within a short period (may happen several times in one year)
Occasional	3	Probably will occur (may happen several times in 1 to 2 years)
Uncommon	2	Possible to occur (may happen sometime in 2 to 5 years)
Remote	1	Unlikely to occur (may happen sometime in 5 to 30 years)

FIGURE 4.3   Sample probability rating scale—NCPS.

Poor	4	Current systems are unlikely to detect occurrence prior to Major or Catastrophic harm to the patient
Fair	3	Current systems are capable of detecting occurrence, however the patient may still incur moderate harm
Good	2	Current systems are likely to detect occurrence with minor risk of harm to the patient
Excellent	1	Current systems are likely to detect occurrence without harm to the patient

FIGURE 4.4   Sample detectability ratings scale.

Like the other parameters, the team must establish acceptable rating systems for weighting how "detectable" these events are in the current system. The following is a sample used by some analysts we have dealt with in the past (see Figure 4.4):

- **Poor:** Current systems are unlikely to detect occurrence prior to major or catastrophic harm to the patient.
- **Fair:** Current systems are capable of detecting occurrence; however, the patient may still incur moderate harm.
- **Good:** Current systems are likely to detect occurrence with minor risk of harm to the patient.
- **Excellent:** Current systems are likely to detect occurrence without harm to the patient.

STEP 7: COMPLETE BASIC FMEA SPREADSHEET

This is the point where there is the most variability on approaches. Some include the detectability column, some do not. Some use the scales above as they are, some modify them for their facilities. This does not mean that any one approach is the correct one or "perfect" one. As mentioned earlier, the ends are more important than the means. Patient safety should remain the focus of such analyses, as opposed to only strict regulatory compliance. Figure 4.5 is a sample spreadsheet for an FMEA using the PROACT Suite software tool mentioned earlier.

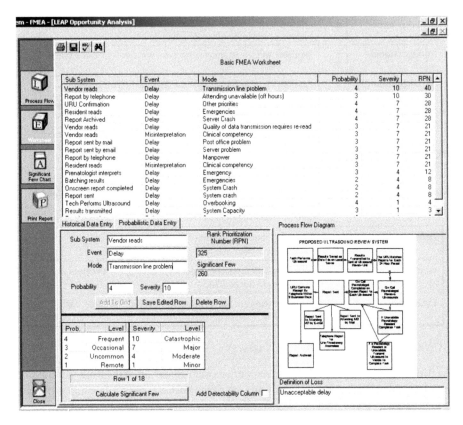

FIGURE 4.5   Sample FMEA worksheet.

DEFINE SUBSYSTEM

Let's take a look at how to fill in the blanks of this spreadsheet. We will start with the first column, which is subsystem. When we created our process flow diagram (PFD), it represented the flow of a "system." Each of the subprocesses within this system (blocks) represents a subsystem. Therefore, each block in the PFD represents a subsystem.

DEFINE EVENT

Move on to column 2, labeled "Event." The event would be consistent with a materialized risk based on our definition of a risk, loss, or failure within the system. For instance, in our proposed ultrasound review system, we were looking for "any event or condition resulting in an unacceptable delay that increases the risk of harm to the patient." The end result here, or the event, is "patient is harmed." Under this scenario, we would put "patient harmed" as the event.

DEFINE MODE

Column 3 is labeled "Mode." Mode has a direct correlation to Event. This is a cause-and-effect relationship. When the Mode occurs (cause), the Event is the effect. If the attending physician was not available (Mode) and therefore there was an excessive delay in treatment for the patient, the effect could be the patient is harmed (Event). The Mode would cause the Event to occur. Any single Event can have numerous Modes. In our example above, there could also be a clinical competency issue (Mode) that resulted in harm to the patient (Event).

We must remember when conducting this type of probabilistic analysis that this is not the same as conducting a root cause analysis (RCA). Contrary to popular belief, we should not expect to be able to determine root causes from an FMEA. Modes are not root causes, as we will learn in later chapters when discussing RCA in detail. Modes are very high-level manifestations of failure that require extensive further drilling down to be able to uncover physical, human, and latent root causes.

Many people get confused about when to apply FMEA, OA, and RCA. Just remember that FMEA and OA are tools to be applied on a system or process, whereas RCA is applied on individual events that occur within these processes. When trying to decide whether to apply an FMEA or OA, we should ask ourselves, "What is our objective for conducting the analysis?" If our objective is to identify potential risks (weaknesses) in our system and to quantify those risks, then an FMEA would be the appropriate tool. If our objective is to identify actual losses incurred with our system, and to quantify those losses, then an OA would be the appropriate tool.

We notice in this approach that the three parameters of probability (P), severity (S), and detectability (D) all have an equal weight as the equation is $P \times S \times D$. The same is true for when we use only Probability and Severity ($P \times S$). This is referenced because when we describe the Opportunity Analysis (OA) tool in the next chapter, we will learn that, when measuring physical losses of event that have occurred, the distributions of dollars lost are not applied so evenly.

STEP 8: DEVELOP CORRECTIVE ACTION PLAN

The goal of the description of FMEA in this text is not to dictate that any particular variation should be used. That the outcomes are the most important factors to consider. If the approach used is capable of producing thorough and credible results, then we should go with it. Any of these variations will still allow for their complementary use with true RCA approaches.

The NCPS approach advocates the use of a hazard-scoring matrix similar to prioritization matrixes used in industry HAZOP (hazardous operations) studies. When the appropriate value is located on the table, it is applied to a decision tree where a series of questions will be asked to determine the recommended actions.

No effort is fail-safe. There is always some degree of risk. Therefore at some point an organization must establish what the minimal level of acceptable risk is. Any identified risk higher than that level will have to be analyzed for ways in which to either reduce the risk to an acceptable level or eliminate the risk altogether.

The 80/20 rule is usually applied in these situations, where the basic FMEA spreadsheet would be sorted to identify the 20% or less of the events found to be accountable for 80% or more of the risk. The 80/20 rule will be described in detail in Chapter 9. A sample outcome of this process may look something like Figure 4.6.

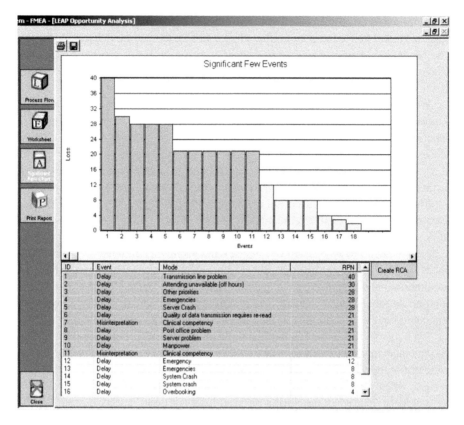

FIGURE 4.6   Sample pareto.

Based on the unacceptable risks identified, action plans will now have to be developed. The NCPS uses the options of Eliminate, Control, or Accept. This is a good basis for moving forward.

ELIMINATE RISK OPTION

To eliminate the risk, a prospective RCA can be done. This will be discussed soon, but basically it treats the event as if it did occur, and then an RCA is conducted to look at how such an event *could* occur. As a result of this prospective RCA, potential root causes will be identified. By addressing these root causes with effective recommendations, the analyst will be considerably reducing the risk of the event occurring again.

CONTROL RISK OPTION

Controlling risks usually will involve developing a plan to minimize the consequences of the event if it were to occur. This may involve using tools like Six Sigma to minimize the process variation caused by such an event. This also may involve implementing detection mechanisms to ensure that the patients themselves are at minimal risk if the event were to occur.

Acceptable Risk Option

This option is usually for events that have been deemed acceptable risks where the patient is unlikely to be harmed if the event were to occur. In these cases, organizations may choose to do nothing and accept the risk identified. Keep in mind the inherent dangers associated with labeling something as an acceptable risk. By documenting these conclusions, it may expose the organization to legal liabilities. That is why being as accurate as possible on these FMEA ratings is so important. NASA described the known O-ring system deficiencies as an "acceptable risk" in the final Challenger commission report. Using an extreme situation such as this shows the possibilities.

Like any action plan, responsibility and accountability should be outlined in the form of task assignments. Timelines should be applied to when the actions will be taken and *metrics to track* should be developed and monitored.

As mentioned earlier, we must reflect on a macro level when trying to decide on our action plans. We should stop and ask ourselves, "How do we define success for this FMEA?" If our primary goal is simply to pass a regulatory audit, our action plans could be remedial, as we know what it will take to "get by." Under these circumstances, we normally understand that our path forward is not comprehensive enough, but time pressures us into taking shortcuts.

If we focus on the patient and ask the same question, our action plans will tend to be much more comprehensive, as we do whatever it takes to prevent unacceptable risks from materializing. This is often a tradeoff between safety and compliance, but nonetheless we are put in these positions every day to make such judgment calls. Making the right call at the right time becomes our challenge!

5 Opportunity Analysis (OA)
The Modified Approach

With all the noise and distraction of a reactive work environment, it is sometimes easy to overlook the obvious. For instance, if one wanted to perform a root cause analysis (RCA) on an event, would they know which event was the most significant or costly? Experience demonstrates that our reactive cultures condition us for focusing on what is happening now (micro) and shield us from looking at the big picture (macro).

This is where the need for a technique to help us take a global look at our situation and assess *all* of the events and their individual impact on overall performance comes into play. There is a technique that will help do just that. It is called *opportunity analysis (OA)*, formerly referred to as *modified FMEA* by Reliability Center, Inc.

Why would one want to perform an OA in the first place? There really are two basic reasons to perform an OA in healthcare. The first and foremost is to make a legitimate business case to analyze one event versus another. In other words, it creates the financial or business case to show a listing of all the events within a given process or system and delineate in dollars and cents (or in weights) why someone would choose one issue from another. It allows the clinical personnel to speak in the language of business.

The second compelling reason is to focus the organization on what the most significant events are, so that quantum leaps in safety and productivity can be made with fewer of the organization's resources being utilized. Experience has shown that the Pareto principle* works with such events just like it does in other areas. It goes something like this: 20% or fewer of the undesirable events that are uncovered by an in-depth OA will represent approximately 80% of the losses for that process being analyzed. This is referred to as 80/20 rule. The 80/20 rules will be discussed later in this chapter.

Let's take a look at a simple example of an OA. One of the first steps involved is to identify the system to be analyzed and draw a basic process flow diagram (block diagram) of the subsystems. In this case the system chosen is a simplified medication order process (Figure 5.1). Next the analyst must define what it is he or she is looking for. For instance, what types of losses are being sought for in this system? In this case the definition will be "any event or condition resulting in a reported near miss over the past 12 months." From there the analysis would look at each of the subsystems and determine what failures have occurred. The OA spreadsheet might look like Figure 5.2.

---

* RCFA Methods course, Reliability Center, Inc., Hopewell, VA, 1985–1998.

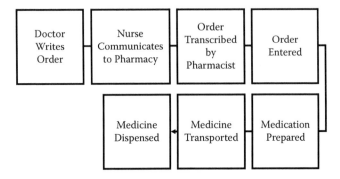

FIGURE 5.1   Sample medication order process—block diagram.

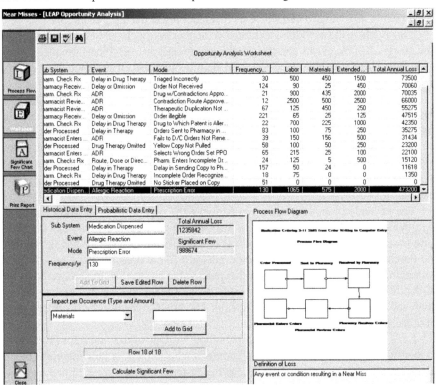

FIGURE 5.2   Medication order process—OA spreadsheet sample.

The idea is to delineate the events that have occurred in the medication order process that have resulted in a reported near miss. In this case, one of the events would be an allergic reaction. The mode of this particular event is that a prescription error was made by the physician. This type of error occurred in this facility 130 times over the past 12 months. The total approximate impact for each occurrence is $3,640 in direct and indirect costs. Now the "frequency" is multiplied by the "impact" for each occurrence, and it is concluded that allergic reactions due to prescription errors are costing the organization approximately $473,200 per year (see Figure 5.2A).

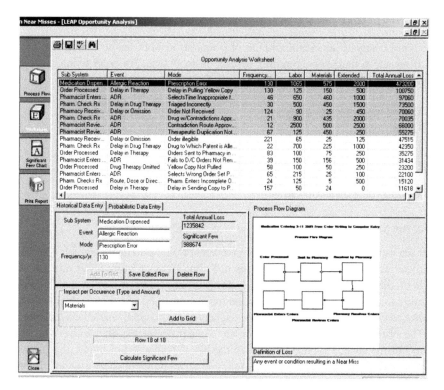

FIGURE 5.2A   Medication order process—prescription errors.

If this analysis were continued, each of the subsystems would be pursued, delineating all of the events and modes that have caused a near miss in their respective subsystems. The end result would be a listing of all the items that have contributed to near misses and hence, their respective losses. Based on the listing, the events identified as the greatest contributors to the annual losses would be candidates for a disciplined RCA to determine the root causes for their existence. At this point, the business case has been made because the events' annual impact has been quantified and the return on investment (ROI) and payback period can be reasonably estimated.

Now that the overall concept of OA is understood, let's take a detailed look at the steps involved in conducting an OA. There are seven basic steps involved:

1. Perform preparatory work
2. Collect the data
3. Summarize and encode results
4. Calculate loss
5. Determine the "significant few"
6. Validate results
7. Issue a report

STEP 1: PERFORM PREPARATORY WORK

As with any analysis, a certain amount of preparation work has to take place. OA is no different, in that it also requires several up-front tasks. In order to adequately prepare to perform an OA, one must accomplish the following tasks:

- Define the system to analyze
- Define loss
- Draw process flow diagram (block diagram)
- Describe the function of each block
- Calculate the "GAP"
- Develop data collection strategy

DEFINE THE SYSTEM TO ANALYZE

Before we can begin generating a list of problems, we have to decide which systems to analyze. This may sound like a simple task, but it does require a fair amount of thought on the analyst's part. Often, analysts try to take an entire facility and make it the system to analyze. This is a prescription for disaster. Trying to delineate all of the failures and/or problems in a moderate-sized acute care hospital would be a daunting task. The system needs to be localized down to one process or system within a larger system. For instance, a typical acute care facility is composed of many disciplines and departments such as ED, OB/GYN unit, pediatrics unit, oncology unit, etc. Even these "systems" may be too large and have to be broken down further into smaller, more manageable systems.

The prudent thing would be to select one unit or process therein at a time and make that system the focus of the OA. For example, in the OB/GYN department, the system selected might be the ultrasound review process, which would have its own subsystems. In other words, the analyst should not bite off more than he can chew when selecting a system to study.

DEFINE LOSS

This may sound a little silly, but the definition of a "loss" must be determined for the system being analyzed. In courses I have taught over the past three decades, students were asked to write down their definition of a loss at their facility. Just about every time, every student has a unique and different definition. The fact is, if event data is going to be collected, everyone involved must be using a consistent definition. If data is being collected and there is no standardized definition, each will provide her own perception of what losses are occurring in her work area. For instance, if a nurse is asked about observed losses in the OB/GYN unit, the list would be completely different from the same list provided by the perinatologist. The dilemma here is that focus is lost when there is not a common definition of what a loss is.

The key to making an effective definition of a loss is to make sure the definition coincides with a particular business need. Here are some definitions that have been used in the past that can be used as examples.

A "loss" is any event or condition:

- Resulting in a reportable sentinel event
- Resulting in a near miss
- Resulting in an adverse drug event (ADE)
- Resulting in an excessive time delay (quantify how long a delay is excessive)
- Resulting in patient dissatisfaction
- Resulting in an unexpected cost in excess of $XX

The definition of a loss is the focal point of the OA. This is setting the parameters around what is going to be considered a "loss" for this analysis. As can be seen in the above list, certain events will be included in one definition and filtered out in another. For instance, a patient may have suffered a mild allergic reaction to a medication because of a prescription error. This event may be included in the ADE and/or near miss definitions, but not the reportable sentinel event definition.

The ED may experience excessive delays in treating patients, but that is not a sentinel event. The definition of loss is like a filter that allows the events that we are seeking to pass through, while blocking the ones that are not the focus of the analysis.

So why bother with a definition? It serves multiple purposes. First, as mentioned above, it is the focal point of the OA. However, the biggest advantage of an agreed-upon definition is that it fosters precise communication between everyone in the facility. It gets people focused on the most important issues.

When a definition of a loss is developed, it should be short and to the point. It is not recommended to have a definition that is several paragraphs long. A good definition can and should be about one sentence long. The definition should only address one business need at a time. For example, a definition that states, "A loss event is anything that causes patient harm, excessive risk, inadequate quality, and excessive costs," is trying to capture too many items at one time, which will cause the analysis to lose focus. If there is a need to look at each of those issues, then the analyst should perform separate analyses for each of the defined losses. It may take a little longer, but the integrity of the analysis focus will be maintained.

Last but not least, it is important to get decision makers involved in the process. It is recommended that someone in authority sign off on the definition to give it some clout. The person in authority may even modify the definition. This will in essence create buy-in from that person and the definition will be *both* of yours and not just yours alone.

Draw Process Flow Diagram (Block Diagram)

A simple flow diagram of the system being analyzed must be developed now that the system has been identified and the definition of a loss determined. This diagram will serve as a job aid later when data collection begins. The idea of a flow diagram is to show the process flow point A to point B. All of the subsystems in that process are to be identified with simple blocks (Figure 5.3).

The purpose of a process flow diagram is to map out a process flow in as simplistic and expressional a way possible. Traditional block diagrams and legends exist where different shapes represent different tasks. If these tools are available, then they should be used, providing they are easy to navigate and easy to understand by those

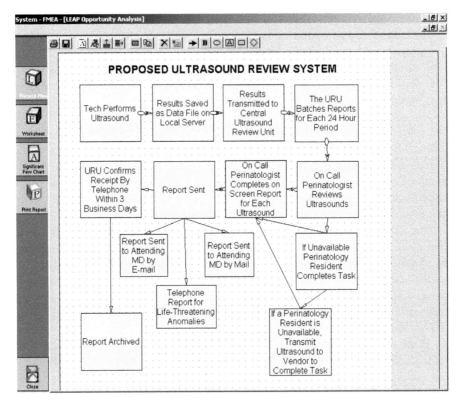

FIGURE 5.3   Current ultrasound review process—process flow diagram.

who will read the diagram. Are they necessary? No. It is insignificant whether all the blocks are the same shape or not. What is important is that the process flow on paper represents accurately the process flow in the field.

DESCRIBE THE FUNCTION OF EACH BLOCK

In some cases, simply drawing the block diagram is not enough of an explanation. Sometimes individuals who (a) are not intimately aware of the function of each of the subsystems, or (b) may be on the OA team will need more explanation. In these cases, it will be necessary to add some level of explanation for each of the blocks. This will allow those who are less knowledgeable in the process to participate with some degree of background.

STEP 2: DEVELOP DATA COLLECTION STRATEGY

The last step in the preparatory stage is to design an interview sheet that is adequate to collect the data consistent with the loss definition and to set up a schedule of people to interview to get the required data. In every analysis the following data are required:

- Subsystem—This correlates to the blocks in the block diagram.
- Event—The event is the actual undesirable event that matches the definition of loss created earlier.
- Mode—The mode is the apparent reason the undesirable event exists.
- Frequency per year—This number corresponds to the number of times the mode actually occurred in a year's time.
- Impact per occurrence—This figure represents the actual cost or weighted value of the mode when it occurs. For instance, impact may comprise materials/supplies $$, labor $$, extended lengths of stay $$, claims $$, etc. This data element can represent any item that has a determinable cost. "Impacts per occurrence" can also be represented using a nominal scale with weights (if dollars are not preferred to be used). For example, if the analyst chose to use "severity" as an impact column, then the following weighted scale may be employed:

  - *Category I:* Catastrophic event (10)—A failure that may cause death or system loss.
  - *Category II:* Major event (7)—A failure that may cause a high degree of customer dissatisfaction.
  - *Category III:* Moderate event (4)—Failure can be overcome with modifications to the process or product, but there is minor performance loss.
  - *Category IV:* Minor event (1)—Failure will not be noticeable to the customer and would not affect delivery of the service or product.*

- The above rating scales will vary based on the system being analyzed and the analysis tool being used. The TJC taxonomy and the NC MERP classifications are two such rating scales.
- Total loss per year—This is the total loss per year for each mode. It is calculated by simply multiplying the frequency per year by the total impact per occurrence.

An effective interview sheet should be based on the definition of loss. The first four columns (subsystem, event, mode, and frequency) are almost always the same. The impact column(s), however, can be expanded to include whatever elements are deemed appropriate for the given situation. For instance, some analysts do not prefer to include straight labor costs since they have to pay such a cost regardless. However, any overtime costs or costs associated for using contract labor may be included if they were associated with the mode occurring.

The last item in the preparatory stage is to create a preliminary interview sheet to list all of the individuals to talk to in order to collect this event information.

---

* NCPS, *The Basics of Healthcare FMEA (HFMEA),* 2004, available at http://www.va.gov/ncps/safety-topics/HFMEAIntro.pdf.

COLLECT THE DATA: PEOPLE OR DATA SYSTEMS—
WHICH DATA IS MOST RELIABLE?

There are a couple of schools of thought when it comes to how to collect the data necessary to perform an OA. On one side, there is the school that believes all data can be retrieved from a computerized system within our organization. The other side believes it would be virtually impossible to get the required data from an internal computer system, since the data going into the system are suspect at best. Both sides are to some degree correct. Data systems do not always give the precise information that is necessary, but they can be useful to verify trends that would be uncovered by interviewing people.

Both of these alternatives will be explored in this chapter and the next. However, the analyst leading the OA will ultimately be responsible for making the decision as to whether the more accurate and timely data will come from interviews or from the existing information systems.

It is recommended that, when using the manual method of data collection (interviewing technique), a two-track approach be taken. Begin collecting data from people through the use of interviews. Once the data have been collected and summarized from the interviews, existing data systems can be used to verify reports and financial numbers and to see if the computer data support the trends that were uncovered in the interviews. The numbers will likely not be the same, but the trends may be. This may also divulge deficiencies in data system capabilities, as many of events uncovered in the interviews will not be found in the electronic recording system.

COLLECTING DATA VIA INTERVIEWS/MEETINGS

Remember from a previous discussion, two job aids were developed. First there was a definition of loss, and second a process flow diagram of the system being analyzed. These two documents will now be used to help structure an interview session. The interviewer would begin by asking the interviewee(s) to delineate any events that meet our definition of a loss within a certain subsystem of the process flow diagram.

This creates a focused interview session. The word "interview" can have a negative connotation. For instance, in order to gain employment, one likely had to go through an interview, which can be stressful. Another negative connotation may be where TV police programs often show a suspect being interviewed (i.e., interrogated) in a dark, smoky room. To avoid any possible negative connotation, these interviews might be referred to as "meetings." Think of such meetings as information-gathering sessions instead of a formal interview. This will certainly improve the flow of information.

Who would be good candidates to include in this opportunity analysis (OA) meeting? It is important to make sure that a good cross-section of people is involved. For instance, it would not be advisable to talk to just nursing staff, because that would be only one perspective. It is best to interview across disciplines, meaning that information is gathered from physicians, lab technicians, pharmacy, facilities, procurement personnel, and administrative personnel. These are a few possibilities. The actual people sought would be those who work at the sharp end, or those closest to the system being analyzed. They would have the best direct experience with the system and with events that have occurred within that system.

It is advantageous when collecting event information to talk to multiple people simultaneously. While this is not always possible, it is preferable. This has several benefits. When one person is talking, it may spur something in someone else's mind. It also has a psychological effect. When people are asked about event information, it may be perceived as a "witch hunt." In other words, they might feel like management is trying to blame people for the event. By including multiple people in the data gathering session, it appears to be more of a brainstorming session than an interrogation.

The interviewing process is really more of an art form than a science. When people first learn to interview, they soon find out that it can sometimes be a difficult task. But like golf or any other type of skill—the more practice with proper technique, the better the final results will be.

An interview is nothing more than getting information from one individual to another as clearly and accurately as possible. To that end, here are some suggestions that will help one become a more effective interviewer. Some of these are very specific to the modified OA process, but others are generic and can be applied to any type of interviewing session.

- *Ask the exact same lead questions of each person.* This will eliminate the possibility of having different answers depending on the interpretation of the question. Later, the facilitator can expand on the questions if further clarification is necessary. The facilitator can use the definition of loss and process flow diagram to keep the session focused on the analysis.
- *Make sure that the participants know what an opportunity analysis (OA) is, as well as the purpose and structure of the data gathering session.* If the facilitator is not careful, the process could begin to feel more like an interrogation than a data-gathering session to the participants. An excellent way to make the participants comfortable with the process is to conduct the sessions in their work environments (i.e., meeting rooms on the floors) instead of in the executive boardrooms. People will be more forthcoming if they are where they are most comfortable.
- *Allow the interviewees to see what the facilitator is writing.* This will set them at ease since they can see that the information they are providing is being recorded accurately. *Never* use a tape recorder in an OA session, because it tends to make people uncomfortable and less likely to share information. Remember, this is an information gathering session and not an interrogation.
- *If the facilitator does not understand what someone is saying, let the person use a pen to draw a simple diagram of the event to explain it.* If it is still not clear, then the facilitator should go out to the actual work area where the problem is to see the area for himself.
- *Never argue with an interviewee.* Even if the facilitator does not agree with the participant, it is best to accept what the interviewee is saying at face value and double-check it with the information from other sources later. The minute the facilitator becomes argumentative, it reduces the amount of information obtainable from the participants.

- *Know the participants' names.* There is nothing sweeter to people's ears than the sound of their own name (as we remember from Dale Carnegie). A facilitator who has trouble remembering names should write the names down for quick reference. This gives any data-gathering session or discussion a more personal feel.
- *Develop a strategy to draw out quiet participants.* There are many quiet people in our workforce who have a wealth of data to share but are not comfortable sharing it with others. The facilitator must be sure to draw out these quiet interviewees in a gentle and inquiring manner. The facilitator can use a nominal group technique where all of the people are asked to write their comments down on an index card and then compile the list on a flip chart. This gives everyone the same chance to have their comments heard, even if they do not wish to speak up in a group.
- *Be aware of body language in interviewees.* There is an entire science behind body language. It is not important that the facilitator become an expert in this area. However, it is important to know that a substantial portion of human communication is through body language, and the facilitator should be aware of some basic facts. For instance, if someone sits back in a chair with arms firmly crossed, this person may be apprehensive and not feel comfortable providing the information that we are asking for. This should be a clue to alter our questioning technique to make that person more comfortable with the situation.
- *Make a note of the extraordinary contributors so that they can assist us later in the analysis.* In any set of interviews, there will be a number of people who are able to contribute more to the process than others. They will be extremely helpful if additional event information is needed for validating the completed OA, as well as assisting us when beginning the actual root cause analysis (RCA).
- *Try to never exceed one-hour data-gathering sessions.* This process can be very intensive, and people can become tired and sometimes lose their focus. This is dangerous because it can upset the validity of the data. So as a rule, one hour of data gathering per interview session is plenty. If more is absolutely required, provide a 15-minute break.

COLLECTING DATA FROM CURRENT DATA SYSTEMS

As we have shown, the effort to collect data straight from the source can be time consuming. However, the data gathered from direct interviews are often the most current and the most accurate in the organization. Facilitators of OA will have the final say as to whether they get the highest data integrity directly from the people closest to the work, or their data collection systems like risk management, incident management, and claims management databases. Accessibility to data in such electronic systems is another issue, as well as getting the data in the formats that you need. These should all be taken into consideration when deciding which the best data source is for the OA.

Many healthcare industry professionals do not believe that current data systems accurately reflect the true activity at the sharp end. Until this confidence is raised, while more cumbersome, the data collected from people are the most reliable.

Understanding why we have a lack of integrity of data could be explored via RCA as well.

STEP 3: SUMMARIZE AND ENCODE DATA

At this stage, a vast amount of data has been generated from interviews. Now the data must be summarized for accuracy. Chances are that when conducting interviews, some redundant data will be collected. For instance, a person from the night shift might be relating the same events the day shift person reported to us. Efforts should be taken to summarize the information and encode it properly so that redundant events, or "double dipping" of events, will not occur.

The easiest way to collect and summarize the data is to input it into an electronic spreadsheet or database like Microsoft® Excel and Microsoft Access. Of course, this could be done manually with a pencil and paper, but if a computer is available, it would certainly be helpful from an administrative standpoint and will save many hours of frustration over performing the analysis manually.

Once all of the information has been input to a spreadsheet, efforts must be taken to locate any redundancy. When inputting information into a spreadsheet, be sure to use a logical coding system. Once a coding system has been defined, stick to it. Otherwise the spreadsheet program will be unable to provide comprehensive results. Table 5.1 shows what is meant by the term "logical coding." If coding portrayed on the bottom of the graphic were used, the attempt to logically sort the data would get erratic results. Therefore, using a coding system like the one depicted in the top of the graphic would provide the required result when using various sorting schemes.

TABLE 5.1
Logical and Illogical Coding Example

Subsystem	Failure Event	Failure Mode
ED	ADE	Prescr. Error
ED	ADE	Dispense Error
ED	ADE	Admin. Error

Logical Coding

Subsystem	Failure Event	Failure Mode
ER	ADE	Presc. Error
Emerg. Rm.	Allergic Reaction	Dispensing Error
ED	ADR	Admin. Error

Illogical Coding

Keep in mind that these tasks being described are those that would be conducted by the OA facilitator *after* all the data have been collected from the participants. At this point, the data are being "cleaned up" for inclusion in the final report. The tasks of summarizing and encoding data are attempts to ensure integrity of the data and to defend the credibility of the conclusions in the final analysis.

How can the facilitator eliminate the redundant information that is given in the interview sessions? The easiest way is to take the raw data from the interviews and input them into the spreadsheet program. From there the powerful sorting capability of the program can be used to help look for redundancies.

The first step is to sort the entire list by the subsystem column. Then within each subsystem, the failure event column will need to be sorted. This will group all of the events from a particular area so that they can easily be recognized for duplicate events. Once again, if logical coding is not used, this will not be as effective. The facilitator should strive to be disciplined in how he or she enters data.

Table 5.2 is an example of how to summarize and encode events. In this example, the ED Subsystem is being reviewed (primary sort) and Adverse Drug Events (ADE) that had occurred (secondary sort). Four different people from four separate shifts gave the facilitator these events. Is there any redundancy? The easiest way to do this is to look at the modes. In this case two modes mention the word "prescription." The second is "order wrong." The interviewee was probably trying to help out by giving the cause of the event, but he or she was also talking specifically about the "prescription error." So in essence the first three events are likely the same event. These three events will have to be summarized into one. Table 5.3 is what it might look like after the events are summarized.

STEP 4: CALCULATE LOSS

Calculating the individual modes is a relatively simple process. Multiply the frequency per year times the total impact per occurrence. If a mode costs $5,000 per occurrence, and it happens once per month (12 per year), then there is a $60,000-per-year problem.

Dollars are typically used when trying to make the business case to the finance people. More than likely, such people will be internal to the organization. Under these circumstances, the finance people will be able to easily see where the losses are distributed throughout the system, right down to the individual events. Dollars

---

TABLE 5.2

Example of Summarizing and Encoding Results

Sub-System	Event	Mode	Frequency	Ext. Stay
ED	ADE	Prescription Error	12	12 Hours
ED	ADE	Wrong Prescription	6	1 Day
ED	ADE	Dr. Order Wrong	12	12 Hours
ED	ADE	Overmedicated	1	5 Days

TABLE 5.3
Example of Merging Like Events

Sub-System	Event	Mode	Frequency	Ext. Stay
ED	ADE	Prescription Issue	12	12 Hours
ED	ADE	Dosage Issue	1	5 Days

are the language of business and are usually the easiest to communicate to all levels of the organization.

However, especially in the healthcare environment, it may be preferable not to use dollars if the reports are intended to leave the organization for review. In such cases, nominal scales or weighting systems may be employed (as discussed earlier). This way it will not be perceived that healthcare decisions were being made based solely on dollars.

Table 5.4 shows a few examples of calculating loss. In this example, frequency per year is multiplied by the impact per occurrence, which in this case is "cost." In other words, when each of these modes occurs, the impact is conveyed in dollars. The last column shows total loss in dollars. Simply multiply the number of lost units by the cost of each unit to give a total loss in dollars. That's all there is to it!

STEP 5: DETERMINE THE "SIGNIFICANT FEW"

Which events, out of all those listed, are significant? We have previously discussed the 80/20 rule, but what does it really mean? This rule is sometimes referred to as the Pareto principle. Pareto was an early 20th century Italian economist who claimed that in any set or collection of objects, ideas, people, and events, a *few* within those sets or collections are *more significant* than the remaining majority. This principle demonstrates that, in our world, some things are more important than others. Let's look at a few examples of this rule in action with things that anyone can relate to:

- Clothing: We wear 20% or less of the clothes that we have 80% or more of the time.

TABLE 5.4
Example of Calculating the Loss

Event	Mode	Frequency/Yr	Impact/Occurrence	Total Loss $
ADE	Prescription Error	130 (1)	3,640 (2)	$ 473,200
ADE	Dispensing Errors	40	3,640	$ 145,600
Excessive Expenditures	Blood Redraws	10,000	300	$ 3,000,000

Frequency × Impact = Total Loss

(1)–JAMA, 1995 Vol. 274 Jul 05: Systems Analysis of Adverse Drug Events
(2)–JAMA, 1997 Vol. 277 Jan 22/29: The Costs of Adverse Drug Events in Hospitalized Patients
 (avg. between non-preventable and preventable ADE's)

- Management: 20% or less of our subordinates consume 80% or more of our time as managers.
- Tools: We use 20% or less of our tools 80% or more of the time.
- Fishing: We use 20% or less of our lures 80% or more of the time.

While these are simple examples, were any of them off base, based on your experience?

This principle works everywhere. Twenty percent or less of the identified events typically represents 80% or more of the resulting losses. This is truly significant if you think about it. It says that if one *focuses* and eliminates the 20% of the events that represent 80% of the losses, quantum leaps in productivity will be achieved in a relatively short period of time. It just makes sense!

To get the maximum effect, it is always wise to present this information in alternate forms. The use of graphs and charts will help to effectively communicate this information to others around us. Figure 5.4 is a sample bar chart that takes the spreadsheet data above and converts it into a more understandable format.

STEP 6: VALIDATE RESULTS

Although the OA is almost finished, there is still more to accomplish. To ensure that our data integrity is solid, it must be validated.

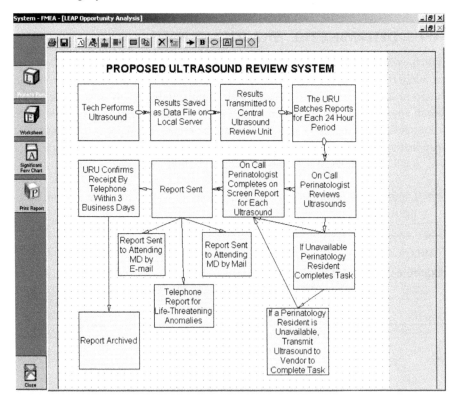

FIGURE 5.4   Sample bar chart of OA results.

At a minimum, the "significant few" events should be double-checked to make sure they are in line with other data sources in the organization. Perfection is not sought in this type of analysis (i.e., down to the penny), simply because it would take too long to accomplish. But efforts should be made to get in the ballpark.

This would be a good opportunity to revisit our more knowledgeable participants to verify trends and financial numbers. If there is ever a controversy over a financial number, numbers that the accounting department deems accurate should be used. Also, it is better to be conservative with financials so that credibility is not lost due to an exaggerated number. The numbers are typically high enough on their own without any exaggeration that this should not be a concern.

Other verification methods might be the existing data system, more interviews, or design of experiments (DOE) in the field to validate data-gathering findings. Efforts should be taken so that these numbers can be presented to anyone in the organization with adequate supporting information to back them up.

STEP 7: ISSUE A REPORT

Last but certainly not least, the findings must be communicated to decision makers so the facilitator can proceed to solve some of these pressing issues. Many people may falter here because they do not take the time to adequately prepare a thorough report and presentation. In order to gain maximum benefit from this analysis, a simple but detailed report must be presented to any and all interested parties (primarily the stakeholders). The report format is based primarily on style. Whether that is one's personal style or a mandated company reporting style is up to the facilitator. The following items are suggested for inclusion in the report in some form.

- *Explain the Analysis.* Many readers of the OA report may not be familiar with the OA process. Therefore, it is in the facilitator's best interest to give them a brief overview of what an OA is and what its goal and benefits are. This way, readers will have a clear understanding of what they are reading.
- *Display Results.* Provide several charts to represent the data uncovered by the analysis. The classic bar chart demonstrated earlier is certainly a minimal requirement. In addition to supporting graphs, all the details should be provided. This includes any and all worksheets compiled in the analysis.
- *Add Something Extra.* This information can be creatively presented to provide further insight into the facility's needs by determining other areas of improvement other than the "significant few." The results could be sorted out in various manners to appeal to individual readers. Such information may shed light on problems in their specific area. Using the querying capabilities of our spreadsheets or databases, any number of interesting insights from this data can be gleaned.
- *Recommend Which Event(s) to Analyze.* Our "significant few" list may reveal any number of events that are "significant." The facilitator cannot work on all of them at once, so the list must be prioritized as to which events should be analyzed first. Common sense would dictate going after the most costly event first. On the surface, this sounds like a good idea, but in reality

it may be better to go after a less significant loss that has a lesser degree of complexity. These are sometimes referred to as the "low-hanging fruit." In other words, first go after the event that gives the greatest amount of pay-back with the least amount of effort.

- *Give Credit Where Credit is Due.* Every person who participated in the OA should be given credit in the report. This includes interviewees, support personnel, etc. If the facilitator wants to gain their support for future analyses, then he will have to gain their confidence by giving them credit for the work they helped to perform. It is also critical to make sure that the results of the analysis are fed back to these people so they can see the final product. A number of analyses have failed because the participants were left out of the feedback loop, so they chose not to help on subsequent projects.

That is all there is to performing a thorough OA. This technique is a powerful analysis tool, but it is also an invaluable sales tool for getting people interested in our projects. It appeals to all parties. The people who participated will benefit because it will help eliminate some of their unnecessary work. Management will like it because it clearly demonstrates what the return on investment will be if those events or problems are resolved. Patients will benefit because they are safer. Keep in mind that it may not always be possible to implement the OA process as cleanly as described in this text. But this overall concept can be taken and applied to a given system so it works best for the analysts.

III

Understanding Root Cause Analysis

6 Understanding Why Things Go Wrong

HOW CAN RCA BE MISPERCEIVED?

Please put aside the industry you work in and follow along from the standpoint of the human being. In order to understand why undesirable outcomes exist, one must understand the mechanics of failure. Most undesirable outcomes are the result of human errors of omission or commission. Experience in industry indicates that any undesirable outcome will have, on average, a series of 10 to 14 cause-and-effect relationships that queue up in a particular pattern for that event to occur.

This dispels the commonly held myth that one error causes the ultimate undesirable outcome. All such adverse events will have roots embedded in the physical, human, and latent areas. These terms will be discussed in detail when we explore the logic tree, but let's summarize them now as an introduction.

Physical roots are typically found soon after errors of commission or omission. They are the first physical consequences of the human decision error. An example of a physical root may be that an incorrect medication was dispensed by the pharmacy.

Human roots are decision errors. These are the actions (or inactions) that trigger the physical roots to surface. An example of a human root or decision error may be that a pharmacist made a unilateral decision to curb the scope of the formulary.

Latent roots are the organizational systems that are flawed. These are the support systems (i.e., procedures, training, incentive systems, purchasing habits, work practices, etc.) that are typically put in place to help our workforce make better decisions. Latent roots are the expressed intent of the outcome. Latent roots are generally the rationale for the decision made. In keeping with our examples above, the reasons the pharmacist made such a decision were (1) a mandate issued by the CFO to cut costs by 10% from each department and (2) a QA/QC system flaw that did not have adequate checks and balances in place to prevent such a unilateral decision from being implemented.

To illustrate this point, Figure 6.1 shows a logic tree diagram (a graphical depiction of cause-and-effect relationships) of an undesirable outcome in healthcare. The event is an *unexpected extended length of stay,* which is a frequent and costly occurrence in healthcare systems. In this particular case, the unexpected length of stay was due to an allergic reaction to a medication (Cephalosporin) given to the patient. Follow each level of the logic tree (Figure 6.2) asking yourself, "How could the previous event have occurred?"

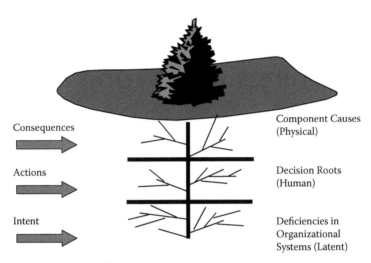

FIGURE 6.1 Description of root systems.

In the end of this analysis, we find that one of the contributors to the undesirable outcome was a CFO's directive to cut costs in all departments by 10% (including the pharmacy). This directive forced the pharmacist to make a conscious decision to reduce the number of versions of the drug to reduce cost. This led to the correct version of the drug not being available when requested. When the drug was requested by the pharmacy, per the doctor's prescription, and entered into the computerized order entry system, it was programmed to default to the least expensive member of that class of antibiotics. The wrong drug was dispensed and ultimately the patient suffered an allergic reaction, resulting in an extended length of stay.

Uses of such disciplined approaches to RCA are evidence based and clearly depict the effect of certain behaviors in an organization and how upper level decision-making can (and does) routinely contribute to undesirable outcomes. Organizations are not doing true RCA if they are not validating their hypotheses with evidence and if they are stopping short of understanding the latent root causes that contributed to the decision errors.

WHAT IS NOT ROOT CAUSE ANALYSIS?

It is common knowledge in healthcare that most all departments are understaffed. When looking further into the increased risk of error by a human being, one should note that being overwhelmed with "emergencies" fuels that environment of error.

Depending on state in which a hospital is located, the task of RCA may fall into the responsibility of quality, risk, or some other performance improvement department. Depending on each state's evidentiary protection laws, RCA will fall into one department or the other. Regardless, the quality and risk managers face significant challenges in accomplishing their objectives in an accurate and reliable manner. Those charged with compliance of applicable regulatory statutes face legal consequences if they do not comply. So compliance is nonnegotiable. The problem is the regulatory requirements for RCA are often minimal, deficient, and superficial when compared to that of the manufacturing industry.

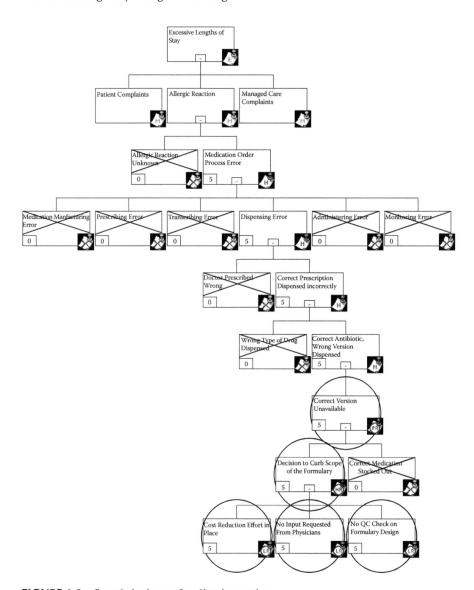

FIGURE 6.2 Sample logic tree for allergic reaction.

One would be astounded to learn of the extent to which the manufacturing industry would go in order to understand why a critical pump failed and consequently shut down an operation, costing millions of dollars in lost product. When compared to the extent that a healthcare facility would generally go to understand why a sentinel event occurred, there is clearly a contrast in the breadth and depth of analysis. It is hard for a healthcare professional to imagine the engineering sciences that go into understanding why equipment fails, because healthcare's requirements are so less intensive in comparison.

When one looks at the parallels between healthcare and industry, the difference is that the pump that healthcare may be working on is the heart! As mentioned in the beginning of this book, this is not because of ill-willed caregivers. Healthcare has a conditioned culture that accepts rudimentary so-called "RCA" methods as being sophisticated enough to be in compliance. Compliance seems to be the primary objective of performing RCAs at the hospital level, as opposed to the potential benefits that can impact real patient safety advances.

Where does this way of thinking come from? The acceptance of common brainstorming techniques such as the Ishikawa Fishbone, the 5 Whys, and process flow mapping have provided healthcare with a false sense of security. This false sense of security comes from the belief that these techniques are comparable to real RCA and therefore used synonymously.

These techniques are referred to as brainstorming techniques, and are not considered RCA techniques within the RCA community. This is because they are not typically based in fact. They allow assumption and hearsay to be viewed as fact. These are attractive techniques to such a reactive environment because they can be concluded very quickly, often in a single session.

Why do such techniques conclude so quickly? Because time is not required to collect data or evidence to support the hearsay (hypotheses). Usually, data collection and testing is the bulk of the time required in any investigation. However, in real investigations, think of what weight data would carry without evidence. If the National Transportation Safety Board (NTSB) didn't collect evidence at airline crash scenes, what credibility would it have when issuing conclusions and recommendations? What weight would prosecutors' cases in court carry if they had no evidence except hearsay?

Another difference between healthcare and industry analysis practices is a moratorium on the time it takes to do the RCA. While industry may set an expected completion date for a critical analysis to be completed, it is understood that the data collection and analysis phases will take as much time as they take. In healthcare, it appears that a 45-day limit is placed to report findings (although extensions are available). Can you imagine the U.S. government telling the NTSB or Columbia Accident Investigation Board that they have 45 days to determine what went wrong? Why don't they do that? Because it is an unreasonable expectation.

In healthcare, the culture itself typically takes the task of RCA lightly. For instance, when RCA meetings are scheduled with designated team members and witnesses, it is common for key individuals not to show up for the meeting. There are usually no repercussions for not attending, so there is no real incentive for participating. The irony here is that many of these RCAs involve unexpected patient deaths. Even more frustrating are the RCA meetings where people speak for the absent people, and this is accepted as fact. Examples are statements such as, "I think what John meant was...." This is a very dangerous habit, and it increases the risk of recurrence of the event, because the true roots may never be explored or uncovered.

When RCAs are accepted in healthcare, they will get full support of executive management in the form of policies and procedures regarding the breadth and depth of RCA. Such in-depth policies and procedures are not commonplace in the manufacturing industry either, but they are becoming more common.

FAILURE MODES AND EFFECTS ANALYSIS (FMEA) VERSUS ROOT CAUSE ANALYSIS

This confusion may have had its foundation from the TJC requirement of RCA being released before the FMEA requirement. From conversations with risk and quality managers, The Joint Commission for the Accreditation for Healthcare Organizations (JCAHO, or "jayco") first had requirements for RCA and outlined suggested formats of what an RCA was and when it would be applied. Then "jayco" issued their FMEA guidelines and focused on high-risk processes. At this point, many people became confused about whether they should do an RCA on a high-risk process or an FMEA on a specific event.

Traditional FMEA concepts came out of the aerospace industry and were developed as a risk tool to aid in the design of new aircraft. They have been adapted and revised for various applications over the years and have now made their way into healthcare. FMEA is a very useful tool in any industry. It is not as complex as it sounds and it can be very versatile in its applications.

A traditional FMEA spreadsheet might look like Figure 6.3. The key focal point is that a system is being analyzed for events that *might* occur in the future. If such events were to occur, what is the "probability" that they would occur, and what

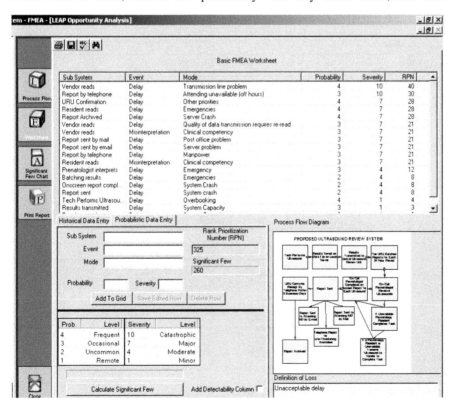

FIGURE 6.3 Sample FMEA spreadsheet of proposed ultrasound review process.

would be the "severity" if they were to occur? Various weighting scales exist both in industry and healthcare. In this method, criteria must be developed before the analysis that would indicate when a proper weight would be applied. Then probability and severity are multiplied, and a risk (or rank) prioritization number (RPN) is arrived at (or a "criticality," as it is often referred to).

This allows us to rank from highest to lowest the events based on RPNs. At this point, a Pareto cut (Figure 6.4) would be made, and the team must decide if any of the events' RPNs are unacceptable and if measures must be taken to reduce the risk of their occurrence. Here the team could employ the RCA method to analyze such high-risk events for their potential causes.

The RCA is employed to focus in on a single event as opposed to the entire process. RCA can be applied to any undesirable outcome. FMEA is merely a technique that provides qualified candidates for RCA.

OPPORTUNITY ANALYSIS (OA) AND RCA

OA is a modification of the traditional FMEA technique discussed above. The difference between an OA and an FMEA is the type of data collected. In the FMEA technique, probabilistic data was used to look at what *might* go wrong.

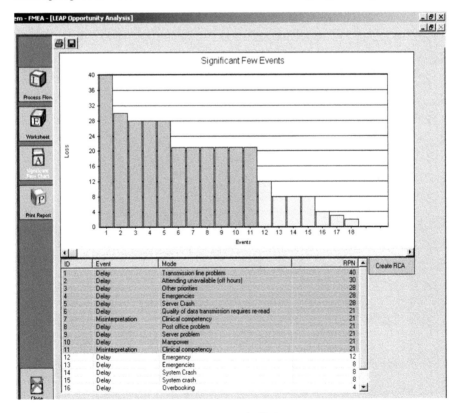

FIGURE 6.4 Sample Pareto cut or "significant few" chart.

OA is the opposite. OA looks at what *did* go wrong. This means utilization of factual and historical data. When using historical data, information about frequency (how often such an event is occurring) and impact (the losses involved when such an event has occurred) is sought. By multiplying frequency times the sum of the total impact, a total annual loss can be calculated.

Like in the FMEA, at this point, a Pareto cut can be made to identify the events that are costing the organization 80% of their losses in the system being analyzed. Again, like FMEA, the critical events are identified that are impacting the current bottom line, and given everything constant, the same failure rate or mean time between failure (MTBF) would be expected (Figure 6.5).

OA is not required by any regulatory agency in healthcare known to the authors; only FMEA is required. However, it is the personal opinion of the author that the need to demonstrate the ROI of patient safety resides in a facility's ability to conduct an OA. This is because in OA, the impacts used for each event can be expressed in dollars *or* in weighted ranking scales (i.e., severity 1 to 4).

Because of the legal concerns in healthcare, the decision to use weighted scales or dollars is dependent on whether the results will be reported to internal personnel or external agencies. Again, evidentiary protection issues control the flow of information.

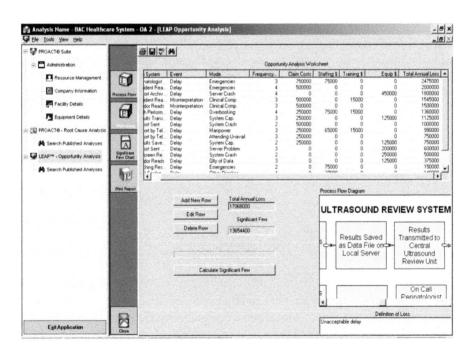

FIGURE 6.5 Sample OA spreadsheet for existing ultrasound review process.

SIX SIGMA AND PROACT RCA

WHAT IS SIX SIGMA AND A BLACK BELT?

Much of what healthcare has come to know about Six Sigma has come from published experiences in industry. Fortune 500 firms such as Motorola, General Electric, AlliedSignal, and DuPont were some of the key industry leaders to put Six Sigma on the map. The intent of this section is to contrast the varying objectives of Six Sigma and PROACT RCA to demonstrate their differences and potential impact on patient safety. Efforts will be taken to show that the two approaches can complement each other if the users are open-minded enough to allow such a "merger."

Polls were conducted on several Internet discussion forums where certified, educated, and experienced people in both Six Sigma and RCA were targeted. First, a basic explanation of what Six Sigma is will be presented, then the contrast with RCA will be made and, finally, the potential impact of these techniques on patient safety will be described.

Taken from a letter in the Greek alphabet, the term "sigma" is used in statistics as a measure of variation. Dr. Mikel J. Harry initially developed and deployed the Six Sigma methodology at Motorola in 1986. In 1993, Dr. Harry teamed with Richard Shroeder to form their Six Sigma academy. Many companies now promote their abilities to implement Six Sigma in their client's corporations.

Breakthrough in profitability is reportedly achieved by reducing direct costs. Direct costs are driven down through Six Sigma projects. Black Belts lead focused projects using the "define, measure, analyze, improve, and control" (DMAIC) approach.

Black Belts have a series of tools in their "toolbox" at their disposal. Some of these primary tools may include the following:

- KISS (keep it simple statistically)
- Cost of poor quality (COPQ)
- Process mapping
- Cause-and-effect diagrams
- Standard operating procedures (SOPs)
- FMEAs
- Six Sigma concepts
- Probability distributions
- Confidence intervals and hypothesis testing
- Identifying a project
- Root cause analysis (RCA)

The following training is typical of that of a Six Sigma Black Belt. This training is often delivered in one-week sessions over a certain time period for a collective training period of four weeks. For instance, for every week of classroom training, there will be three weeks of work in the field utilizing their new learning on their first project. While there are various vendors that provide such training, the range of costs to "certify" a Black Belt is reportedly from $11,000 to $40,000 per person, with the median being in the upper range. The number of Black Belts necessary in a given company will vary from a percentage of the organization to using net profit before income as the basis. Black Belts are typically expected to yield $150,000 per project, but actual reported averages by users tend to indicate that number to be around $75,000 per project. Whether results are above or below reported averages depends on the knowledge and skills of the analysts.

Below is a listing of the typical topics in a Black Belt curriculum along with a brief description.

Statistical Tools

SPC—Statistical Process Charts
- Control
  - Variable Control Chart
  - Attribute Control Chart
- Process Capability
- Statistical Analysis
  - Correlation Analysis
  - Test of Significance
- Design of Experiment (DOE)

Process Mapping Tools

- Flow Charts
- Work Flow Diagrams
- Brown Paper Flow Chart
- Lean Process

Data Display Tools

- Pie Chart and Bar Graphs
- Histogram
- Pareto Analysis
- Scatter Diagram
- Trend Chart
- Concentration Diagram

Problem-Solving Tools

- 8D Method
- Cause and Effect (Fishbone Diagram)
- Brainstorming
- Pert Diagram (Program Evaluation and Review Technique)
- Gantt

Root Cause Analysis Tools

- Time Line Analysis
- Fault Tree Analysis
- 5 Whys
- What Is

Process Improvement Tools

- Failure Mode and Effects Analysis (FMEA)
- Mistake Proofing
- 5 Ss
- Total Production Maintenance (TPM)
- Setup Reduction

Product/Process Interaction Tools

- Quality Function Deployment (QFP)
- Failure Mode and Effects Analysis (FMEA)
- Design For Assembly, Manufacturability, Environmental Friendly (DFA/M/E)
- Design of Experiment (DOE)
- Test of Significance
- Musts and Wants
- Nominal Group Techniques, Voting and Ranking
- T-Tests
- F-Tests

Lean Thinking Tools

- Work Flow Analysis
- Kanbans
- Take Time
- Setup Reduction
- One Piece Flow

This has been a purposeful attempt to elaborate on the extensiveness of Six Sigma Black Belt training and its requirements. In manufacturing industry experience, Six Sigma advocates (vendors) usually sell in at the CEO level for a minimum of $1,000,000 or they do not work with that corporation. This is quite a bit of clout on the vendor's part.

When a CEO signs a check for that amount of money, the commitment trickles down throughout the corporation. For instance, look at Jack Welch, the former CEO of GE who was a big part of putting Six Sigma on the map. He basically said that if you do not agree with the principles of Six Sigma in GE, you do not work for GE. Jack Welch's right-hand man at the time, Larry Bossidy, went on to become the CEO of AlliedSignal (now Honeywell) and that trend was continued. This is a serious commitment, and everyone knows where the CEO sits on this issue. Also with this type of commitment comes success, at least on paper and in the media, even if the data is not there to back it up. On paper, such efforts signed off on by a CEO typically do not fail on paper but may fail in reality. This is not only true of efforts such as Six Sigma but most any effort.

What CEO wants to appear in an article in *Fortune* or *Time* about how his initiatives failed? While relatively few of the Six Sigma successes in industry were popularized as much as the GE, AlliedSignal, or DuPont cases, we do not hear as much about the companies where Six Sigma did not meet expectations, which were many.

These comments are not intended to take anything away from what the Six Sigma concepts are capable of, but to highlight that just because this concept worked in one culture does not mean it will replicate the success in another. This is true for any initiative, anywhere. The key here is the total executive commitment to support the process throughout the entire corporation.

WHAT ARE THE PRIMARY DIFFERENCES BETWEEN SIX SIGMA AND RCA?

Where does RCA fit in Six Sigma? The focal point of almost any Six Sigma effort is to achieve precision through the minimization of process variation. However, the goal of RCA is not to minimize process variation but to eliminate the risk of recurrence of the event that is causing the variation.

For instance, if an ED was the system being analyzed, Six Sigma might seek to minimize the consequences of ADEs by implementing recommendations that would catch any adverse reactions to medications faster in order to take corrective actions and minimize the consequences (i.e., harm to patient, cost to facility, etc.).

Whereas RCA would seek to drill down on the individual ADE incident and understand the chain of events that led to the ADE in the first place, RCA would uncover the system deficiencies that triggered poor decisions being made that set off a series of physical consequences until a patient was affected. *RCA seeks to understand what causes the events to occur, and Six Sigma seeks to minimize the consequences of those events when they do occur.*

Traditionally, Six Sigma toolboxes utilize total productive maintenance/management (TPM) problem solving, brainstorming, and RCA tools such as 5 Whys, Ishikawa Fishbone diagrams, fault tree analysis, and timeline analysis. While these tools are good for basic problem solving, they are not traditionally used to the extent that RCA will be described in this text. As stated earlier, the conceptual goals of the two approaches are distinctly different; therefore, the breadth and depth of the analyses will vary.

SIX SIGMA IN INDUSTRY VERSUS HEALTHCARE: WHAT'S THE DIFFERENCE?

Let's draw the parallels to the implementation of Six Sigma in healthcare, as opposed to industry:

Industry	Healthcare
Six Sigma Black Belts are typically engineers who have a background in statistics.	Six Sigma Black Belts will likely not be engineers or people with a heavy background in statistics.
Six Sigma was developed on the basis of statistical process control/statistical quality control (SPC/SQC). This means that instrumentation was installed on process lines in the form of distributive control systems (DCSs). Such systems would monitor in real time any process variance in accordance with preprogrammed min/max alarm limits.	Healthcare, from my observations, does not have a stable process like in industry. I will use the example of the emergency room where the level of resources required, the type of services required, and the frequency and variability of the patient load are not set. This set of uncertainty would make the application of Six Sigma (as in industry) difficult to apply with controlled results.
Six Sigma was intended to add precision to an existing system that was in "control" and at a high level of reliability (i.e., to move from 99 to 99.9997% when measuring defects per million opportunities (DPMOs).	Such control and high reliability of systems are not the norm in healthcare. Six Sigma was not designed to take a system with poor reliability and bring it up to precision levels.
Six Sigma typically requires on average: Number of projects/site: 7.5 Number of team members/project: 8 Number of meetings/project: 15 Number of hours/meeting: 3 Number of homework hours/person: 17.5 Number of man hours/yr/site: 9840 man hours	From observations in healthcare, they do not have the staffing and/or resources to allocate such time for such an initiative. It is hard enough to get a team together for a single meeting and have everyone show up, much less repeated meetings requiring homework.
Cost for man-hours alone @$25/hr: $246,000/site in man-hours alone (training costs excluded).	Given the lean resources already available in healthcare, my concern would be that patient safety would initially be negatively impacted as valuable resources are forced to spread their focus instead of narrowing it on the patient.

Six Sigma (in concept) has many of the elements that can enable success in organizations. While the tools within Six Sigma are not generally new, the aggregation (expression) of them into the "system" is novel. Like any initiative, the key to the success is in the support, implementation, and sustainability of the effort. The same is true for RCA!

7 The PROACT® Root Cause Analysis (RCA) Methodology

I will start this chapter off with a pet peeve of my own that I want to dispel for you all. When I am reading any text and I see the ® (registered trademark) symbol, it makes me suspicious about the motives of the author. This would insinuate something that is proprietary and that the author is marketing. I do not like paying good money for a book that turns out to be a sales pitch and does not provide me value.

Now let me explain my rationale for using the ® in this chapter. I find it almost impossible to write a text on "root cause analysis" or RCA because there is no universally accepted standard as to what an RCA is. There are no regulations that can mandate that a certain type of RCA be used; only what the desired outcomes (in terms of information) should be. Because of this lack of clarity of what RCA is, practitioners define whatever they are doing to solve problems as RCA. In essence, the standalone term *RCA* is diluted to the point of being useless and misleading. RCA is simply a noun in today's environment.

The only way for people to get value from RCA is to use a specific methodology in which they have had success. These "specific" methodologies represent the various proprietary brands on the market. These brands are simply adjectives of the noun RCA. My firm developed one of these brands, referred to as PROACT®. This is an RCA process that has been used successfully all over the world since 1985. To protect the uniqueness of our brand of RCA, we must formally register it to prevent others from using the same name. So now you know why we have had to become a victim of our own pet peeve!

The term PROACT has recently come to light to mean the opposite of react. This may seem to be in conflict with PROACT's use as a root cause analysis (RCA) tool. From this point on, the term *RCA* shall be used as being consistent with the PROACT RCA methodology, which is simply a brand of RCA. Normally, when one thinks of RCA, the phrase "after the fact" comes to mind. This is because most of the time the undesirable outcome must occur first before we will do something about it. So how can RCA be coined proactive?

In the last chapter, on opportunity analysis (OA), a process was clearly outlined that identified which failures or events were actually worth performing RCA on. This prioritization technique taught us that, generally, the most valuable and important events to analyze are *not* sporadic incidents, but rather the day-to-day chronic events that drain profitability and efficiency.

RCA tools can be used in a reactive and/or a proactive fashion. The analyst will ultimately determine this. When using RCA only to investigate "incidents" that are defined by regulatory agencies, then all efforts are simply responding to reactive, short-term needs of the daily operation. However, if we were to use the OA tools described previously to prioritize their efforts, we will uncover events that many times are not even recorded in the incident management systems (IMS) or equivalent. This is because such events happen so often that they are no longer an anomaly; they are a part of the job. They have been absorbed into the daily routine. By uncovering such events and analyzing them, the organization is being proactive, because it is now looking at such chronic events that were normally overlooked.

The greatest benefits from performing RCAs will come from the analysis of chronic events, hence using RCA in a proactive manner. Often, analysts get sucked into the "paralysis-by-analysis" trap and end up expending too many resources to attack an issue that is relatively unimportant. These are commonly referred to as the "political failures of the day." Trying to do true RCA on everything will destroy a company. It is overkill, and companies do not have the time or resources to do it effectively (see Figure 7.1).

Understanding the difference between chronic and sporadic events will now highlight an awareness to which data collection strategy will be most appropriate for the event being analyzed. The key advantage (if there can be one) with chronic events is their frequency of occurrence. This is an "advantage" because like the detective stalking a serial killer, he or she is looking for a pattern to their activities. In this manner, the detective may be able to stake out where he feels the next logical crime will take place and hopefully prevent its occurrence. The same is true for chronic events. Chronic failures will likely happen again within a certain time frame, and then we may be able to plan for the recurrence and capture more data at that point.

Conversely, when looking at what data collection strategy would be employed on a sporadic event, the frequency factor does *not* work in our favor. Under these circumstances, our detective may be investigating a single homicide and rely only on the evidence at that scene. This would mean the analyst must be diligent about collecting the data from the scene before it is tampered with and possibly tainted. Related to our environments, when a sporadic event occurs, analysts must be diligent at that time to collect the data in spite of the massive efforts to get operations back to normal.

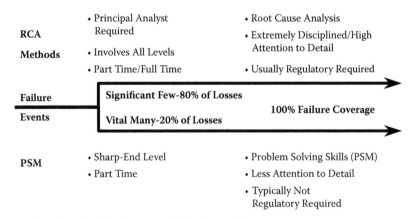

FIGURE 7.1 The two-track approach to failure avoidance.

PRESERVING EVENT DATA

The first step in the PROACT RCA process, as is the case in any investigative or analytical process, is to collect and preserve pertinent data (evidence). Before discussing the specifics of how to collect various forms of data and when to collect it, the psychological side of why people should assist in collecting data from an event scene should be understood.

Consider this scenario: a patient has a cancerous tumor in his lung. A surgeon determines that conducting a laser bronchoscopy can remove the tumor. The patient goes under anesthesia. During the procedure, while the laser is inserted into the bronchoscope, a small but brief fire erupts in the right bronchus. The instruments are quickly removed, and the fire is suppressed. The patient suffers minor injuries and ultimately recovers fully.

What is the likely scenario in this case with regard to data collection efforts? First, the surgeon and his team make every effort to ensure that the patient is safe and stabilized. The patient, who was under anesthesia, does not know what happened. Once the patient is stabilized, efforts are taken to clean up the OR for the next scheduled surgery. What happened to the urgency to collect event data?

What if we were the risk manager (RM) in this case and we were required to investigate all reportable sentinel events? The event would first have to be reported. In this case, if the incident was not reported and the patient was not aware of the incident, then this situation may not be investigated. The reasons behind the decision not to report the incident would be the basis of another book (that would be double the size of this one).

Ideally, this incident should be reported, and it would be considered a sentinel event per the TJC guidelines.* Say this is the case. After the incident, the staff is tending to the patient in the recovery room. The hospital staff is cleaning the OR. The RM wants to collect data from the OR related to the surgery. By the time the RM gets to the OR, it has been cleaned up completely, and nothing from the incident remains.

Even if the RM gets to the OR when the staff is cleaning and requests that they leave for an hour while data is collected, how will they likely respond? Because in this scenario there is no RCA policy or procedure, the people cleaning the OR will consider the RM crazy. They will think, "Doesn't the RM realize the cost of a lost hour in the OR and that the entire schedule will be backed up if we wait an hour?"

When this mentality prevails, it is like a police detective going to a crime scene that has been cleaned up and all the evidence thrown away. That detective must make a "solid case" for court, and how can he do that with no evidence?

In an ideal situation, a formal RCA team would be amassed, and assignments would be made to collect data from the scene immediately. Given the "witch-hunting culture" that commonly exists, the hospital staff may ask "Why should we uncover data/evidence that may incriminate us?" While this is a hypothetical scenario, it could represent many situations in any working environment. What is the incentive to collect event data? This is a time-consuming task, and it will often lead to people who used poor judgment in their decisions; therefore, management could conduct a witch hunt.

---

* TJC, *Sentinel Event Policy and Procedures*, 2004, available at http://www.va.gov/ncps/safetytopics/ HFMEAIntro.pdf.

Subsequently, it could lead to uncovering poor organizational or management systems and poor decisions by physicians, so what could happen to the RM if these things are concluded?

These are all valid concerns. This author has seen the good, the bad, and the ugly created by these concerns. The fact is that, if there is a desire to discover the truth—the real root causes—the analyst cannot do so without the necessary data. Think about any investigative or analytical profession. The first step is always to collect data. Is a detective expected to solve a crime without any evidence or leads? Is an NTSB investigator expected to solve the reasons for an airline accident without any evidence from the scene? Can doctors make diagnoses without any information about the patient's symptoms? If these professionals see the necessity of data and information to draw accurate conclusions, then it must be recognized that there is a similar correlation to RCA.

There is a general resistance to data collection for RCA purposes. We can draw two general conclusions from experience (see Figure 7.2):

1. People are resistant to collecting event data because they do not appreciate the value of the data to an analyst.
2. People are resistant to collecting data because of the paradigms that exist with regard to "witch hunting" and managerial expectations.

The first conclusion above is the minor of the two. Often, production in any facility is the ruling body. Organizations are paid to produce a quality product, whether that product is patient care, oil, steel, or on-time package delivery. When the production mentality is dominant, it forces the organization to react with certain behaviors. If production is paramount then, whenever an event occurs, it must be cleaned up quickly to keep the operations engine going. The focus is not on why the event occurred; rather, it is on the fact that it did occur, and the operation must get back on track.

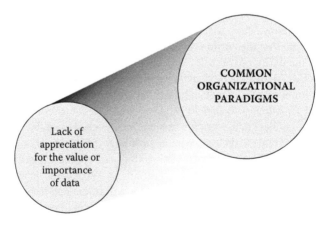

FIGURE 7.2 Typical reasons why event data is not collected.

This paradigm can be overcome with awareness and education. Executive management must first commit to supporting RCA both verbally and on paper. As discussed earlier in Chapter 2, demonstrated actions are what are seen as "walking the talk," and one of those actions was issuing an RCA policy and/or procedure. This would make collection of data a *requirement* rather than an option. Also, it is not enough to just support data collection. Data collectors must clearly understand *why* they should collect the data and *how* to do it properly.

People's responsibilities should be taken into consideration to show them the purpose of data collection and how it will benefit them. If a nurse in a patient room is the first one to an event scene, she should understand what is important versus unimportant information to a potential RCA. For instance, in the bronchoscopy example cited earlier, what key pieces of data should have been collected and preserved? Who should collect them? Do they know they should collect them?

One who understands how important data is to an analysis will appreciate why it should be collected. If people do not understand or appreciate its value, then the task is seen as a burden to an already full plate of things to do. Anyone with basic training in proper data collection procedures can prove invaluable to any organization.

The news will often highlight the potential consequences of poor data collection efforts in a high-profile court case. Allegations are made as to the sloppy handling of evidence in lab work, improper testing procedures, improper labeling, and contaminated samples. Poor data collection methods can cause analysts and/or their counsels to lose their cases (either in or out of court).

Providing the above support and training overcomes one hurdle, but it does not clear the hurdle of perceived witch hunting by an organization. People may choose not to collect data for fear that they may be targeted based on the conclusion drawn from the data. This is a very prominent cultural issue that must be addressed in order to progress with RCA. "Root" causes are difficult to determine if a witch-hunting culture is prevalent.

THE ERROR-CHANGE PHENOMENON

Research, paired with experience, indicates an average number of errors must queue up in a particular pattern for a catastrophic event to occur. The error chain concept,* "...describes human error accidents as the result of a sequence of events that culminate in mishaps. There is seldom an overpowering cause, but rather a number of contributing factors of errors, hence the term *error chain*. Breaking any one link in the chain might potentially break the entire error chain and prevent a mishap." This research comes from the aviation industry and is based on the investigation of more than 30 accidents or incidents. This has been our experience as well in investigating healthcare-related failures.

Flight Safety International states that the fewest links discovered in any one accident was four, the average being seven.† Our experience in industrial and healthcare applications shows the average number of errors that must queue up in a particular

---

* Flight Safety International, Crew Resource Management Workshop, September 1993.
† Flight Safety International, Crew Resource Management Workshop, September 1993.

sequence to be between 10 and 14. This is the core to understanding what an analyst needs in order to determine why undesirable events occur.

The term we will use here for this phenomenon is "error-change relationships." First some terms must be defined in order to communicate more effectively. A modification of James Reason's definition of error* will be used for our RCA purposes. An error will be defined as "an action planned, but not executed according to the plan." This means that we intended on a satisfactory outcome, and it did not occur that way. In some manner, our efforts deviated from our intended path. The change, as a result of an error in our environment, is something that is perceptible to the human senses. An example might be that a nurse administers the wrong medication to a patient, and the adverse reaction is the perceptible change. These series of errors and associated changes occur around us every day. Catastrophic occurrences occur when they queue up in a particular pattern (see Figure 7.3). With this in mind, the following holds true.

1. As human beings, we have the ability through our senses to be aware of our environments. If our senses are sharpened, we can detect these changes and take action to prevent the error chain from running its course. Many of our organizational systems are put in place to recognize these changes. For example, the nurse's station in the intensive care unit (ICU) is full of screens attached to monitoring equipment on patients. This monitoring equipment is in place to identify the various changes in the patient's condition. If these changes are not within acceptable limits, actions are taken to guide them within acceptable limits.

2. By witch hunting the last person associated with an event, the organization gives up the right to the information that person possesses on the other errors that led up to the event. If a person associated with the event is disciplined because the culture requires a "head to roll," then that person (or anyone nearby) will not be honest about why the decisions that resulted in errors were made.

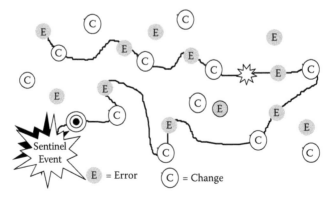

FIGURE 7.3   The error-change phenomenon.

* Reason, James, *Human Error,* Cambridge University Press, New York, 1990.

In Chapter 9, the logic tree will be explored. This is a graphical representation of an error-change chain. This error research by Reasons and Flight Safety International is discussed at this point, because it is necessary to understand that any investigation or analysis cannot be successfully performed without data. Experience dictates that, in the field, application of RCA and the physical activity of obtaining such data can have many organizational barriers in front of it. Once these barriers are recognized and overcome, the task of preserving and collecting the data is relatively simple.

THE "5 PS" CONCEPT

PReserving failure data is the "PR" in PROACT. In a typical high-profile RCA, an immense amount of data is typically collected and then must be organized and managed. This discussion will focus on how to manage the data collection process manually versus with software. Automating the RCA documentation process will be handled in the end of this text.

Reflect back on the fire that occurred during the bronchoscopy procedure mentioned earlier. Assume it was a reportable sentinel event requiring an RCA to be conducted. Also assume the patient was made aware of the incident and has filed a claim against the hospital and the surgeon. We are charged to collect the necessary data for an investigation. What information needs to be collected? We use a 5 Ps approach to guide this process, where the Ps stand for the following:

1. Parts
2. Position
3. People
4. Paper
5. Paradigms

Virtually anything that needs to be collected from an event scene can be categorized under one or more of these headings. Many items will have shades of gray and may fit under two or more headings, but the important thing is to capture the information and slot it under one heading. This categorization process will help manage the data for the analysis.

The parallel of the detective will be used here again. What do detectives and police routinely do at a crime scene? The police typically rope off the area, preserving the positional information. The detectives interview people who may be eyewitnesses. The forensic teams "bag and tag" evidence or parts. There is a hunt for information or a paper trail of a suspect that may involve past arrests, insurance information, financial situation, etc. And lastly, as a result of interviews with observers, tentative conclusions are drawn about the situation such as, "…he was always at home during the day and away at night. We would see children constantly visiting for five minutes at a time. We think he is a drug dealer." These are the paradigms that people have about situations that are important because, if they believe these paradigms, then they base decisions on them. This can be dangerous, but it is reality.

Parts

Parts will generally mean something physical or tangible. The potential list is endless, depending on the industry where the RCA is conducted. The following list gives some examples of *parts* in the healthcare field:

- Medical diagnostic equipment
- Surgical tools
- Gauze
- Fluid samples
- Blood samples
- Biopsies
- Syringes
- Testing equipment
- Needles
- Tissue samples
- IV solutions
- X-rays
- Monitoring strips
- Medication
- Air samples

Position

Position data is the least understood and is one of the most important to an RCA. Positional data comes in the form of two different dimensions, one being physical space and the second being point in time. Positions in terms of space are vitally important to an analysis because of the facts that can be deduced.

When the space shuttle Challenger exploded on January 28, 1986, it was approximately five miles in the air. Film footage shot from the ground provided millisecond-by-millisecond footage of the parts that were being dispersed from the initial cloud. From this positional information, trajectory information was calculated, and search-and-recovery groups were assigned to approximate locations of vital parts. Approximately 93,000 square miles of ocean were involved in the search and recovering of shuttle evidence in the government investigation.* This shows how positional information is used to determine, among other things, force.

Positions in time are extremely important to any RCA. Monitoring "positions" within a reasonable amount of time for undesirable outcomes to occur can provide information for correlation analysis for establishing relevant trends. By recording historical occurrences, plots can be trended that identify the presence of certain variables when these occurrences happen. Let's take a look at the shuttle Challenger again. The conclusion that was reported to the public was an O-ring failure resulting in a leak of solid rocket fuel. Looking at the positional information from the standpoint of time, the O-rings had evidence of secondary O-ring erosion on 15 of the previous 25 shuttle

---

* Video, *Challenger: Disaster and Investigation,* Cannata Communications Corporation, 1987.

launches.* When the solid rocket boosters (SRBs) are released, they are parachuted into the ocean, retrieved, and analyzed for damage. The correlation of these past launches, which incurred secondary O-ring erosion, showed that low temperatures were a variable. The "position in time" information aided in this correlation.

Moving into a more familiar environment, we can review some general positional information that can be collected at most any healthcare organization:

- Physical position of parts at event scene
    - Where were O_2 tanks before the MRI machine activated and drew them into the patient?
- Point in time of current and past occurrences
    - Do the ADEs occur more on a certain shift or rotation?
- Position of personnel at time of occurrence(s)
    - Where was the patient's nurse when the patient coded?
- Position of occurrence in relation to overall facility
    - Do more ADEs occur in pediatrics than the ED?
- Environmental information related to position of occurrence such as temperature, humidity, wind velocity, etc.
    - Was the air quality in the OR a contributor to the nosocomial infection that patient contracted?
- Monitoring screen positions
    - Are the monitors in such a position they are more likely (and easier) to be seen by the ICU nurses?

This is just a sampling to get individuals in the right frame of mind of what is meant by "positional" information (see Figure 7.4).

PEOPLE

The "people" category is the most easily defined "P." This is simply who should be interviewed initially in order to obtain information about an event. The people who must be interviewed first should typically be the physical observers or witnesses to the event. Efforts to obtain such interviews should be relentless and immediate. The analyst will risk the chance of losing the details associated with direct observation if he conducts an interview days after an event occurs. Delayed interviews result in a loss of some degree of short-term memory recollection and also risk the observers having talked to others about their opinion of what happened. Once an observer discusses such an event with an outsider to the situation, the observer will tend to reshape the direct observation with the new perspectives.

The goal of an interview with an observer should be for that the interviewer to see through the interviewee's eyes and clearly see what was observed. The description must be that vivid, and it is up to the interviewer to obtain that clarity through questioning.

---

* Lewis, Richard, *Challenger: The Final Voyage,* Columbia University Press, New York, 1988.

FIGURE 7.4   Human factors design of monitoring stations.

Interviewing skills are necessary in such analytical work. People must feel comfortable around an interviewer and not intimidated. A poor interviewing style, just like a poor bedside manner, can ruin an interview and subsequently an analysis or investigation.

A good interviewer will understand the importance of body language. Experts estimate that approximately 55 to 60% of all communication between people is through body language. Approximately 30% is through the tone of voice, and 10 to 15% is through the spoken word.* This emphasizes the need for an interview to be conducted in person rather than over the telephone when possible. In the legal profession, lawyers are professionals at reading the body language of their clients, their opposition, and the witnesses. Body language clues will direct their next line of questioning. This should be the same for interviews associated with an undesirable event in the healthcare industry. The body language will tell the interviewer when he is getting close to the desired information, and this will direct the line and tone of subsequent questioning.

When interviewing during the course of an RCA, it is important to consider the logistics of the interview. Where is the appropriate place to interview? How many people should be interviewed at a time? What types of people should be in the room at the same time? How will all the information be documented? Preparation and environment are very important factors to consider.

We discussed the interviewing environment and the ideal number of people in an interview in Chapter 5. Those same pointers hold true when interviewing for the actual RCA versus the OA.

Interviews are most successful when the interviewees are from various departments, and more specifically from different "kingdoms." A kingdom can be defined as an entity that builds castles within the facility and tends not to communicate with

---

* Lyle, Jane, *Body Language,* Hamlyn Publishing Group, London, 1990.

others. Examples are nurses versus physicians, facilities versus operations, quality versus risk, administration versus professional clinical staff, etc. When such groups get together, they learn a great deal about the other's perspective and tend to gain respect for each other's position. This happens after the fact, because the first thing that must occur is that all groups participate in the activity. This is another added benefit of an RCA in that people actually meet and communicate with people from different levels and areas.

An interviewer may be fortunate enough to have an associate analyst (assistant) to take notes while the interviewer focuses on the interview itself. It is not recommended that recording devices be used in routine interviews, as they are intimidating, and people might feel that the information may be used against them at a later date. In instances where severe legal liabilities may be at play, legal counsel may record such interviews. However, if this occurs, the counsel is generally doing the interviewing, not an RCA analyst. In the case of most chronic failures or events, use of recording devices is rare.

Typical people to interview will be based on the nature of the organization and the event being analyzed. As a sample of potential interviewees, please consider the following list:

- Eyewitnesses
- Physicians
- Nurses
- Management personnel
- Administration personnel
- Clinicians
- Technical personnel (i.e., lab techs)
- Purchasing personnel
- Supply room personnel
- Vendors/manufacturers' representatives
- Other similar sites with similar processes
- Quality and risk personnel
- Pharmacy personnel
- Patient safety personnel
- Outside experts
- Patient
- Patient's family

This list gives a feel for the variety of people who may provide information about any given event, but by no means is it exhaustive.

PAPER

Paper data is probably the most understood form of data. In this information age where instant access to data through our communications systems has become the norm, a great deal of paper data can be amassed. However, analysts should make sure they are not collecting paper data for the sake of developing a big file. Some organizations seem to feel they are getting paid based on the thickness of the file folder. All efforts should be made to ensure all the data collected are relevant to the event at hand.

Keep in mind the detective scenarios described earlier and the fact that they are always working toward preparing a solid case for court. Paper data is one of the most effective pieces of evidence in court. Solid, organized documentation is the key to a winning strategy.

Typical paper data examples are as follows:

- Patient record
- Physician's orders
- Lab reports
- Equipment specifications
- Pharmacy records
- Literature search results
- Shift logs
- X-rays
- Procedures
- Policies
- Financial reports
- Training records
- Purchasing requisitions/authorizations
- Quality control reports
- Employee file information
- Equipment maintenance histories
- Patient load histories
- Medical histories
- Safety records information
- Internal memos/E-mails
- Labeling of equipment/products
- Monitoring equipment reports
- Past risk assessments
- OR scheduling records

In Chapter 12 we will discuss how to keep all this information organized and properly documented in an efficient and effective manner.

PARADIGMS

Paradigms have been discussed throughout this text as a necessary foundation for understanding how our thought processes affect our problem solving abilities. But exactly what are paradigms? Our definition is based on futurist Joel Barker's:

> A paradigm is a set of rules and regulations that (1) defines boundaries and (2) tells you what to do to be successful within those boundaries. (Success is measured by the problems you solve using these rules and regulations.)*

---

* Barker, Joel, *Discovering the Future: The Business of Paradigms,* ILI Press, Elmo, MN, 1989.

This is basically how individuals view the world and how they react and respond to situations that arise around them. This inherently affects how one approaches solving problems and will ultimately be responsible for success or failure in the RCA effort.

Paradigms are a by-product of interviews carried out in this process as discussed earlier in this chapter. Paradigms are recognizable because repetitive themes are expressed in these interviews from various individuals. How an individual sees the world is a mindset. When a certain population shares the same mindset, it becomes a paradigm. Paradigms are important because, even if false, they represent the beliefs on which decision making is based. Therefore, true paradigms represent reality to the people who possess them.

Below is a list of common paradigms observed in healthcare. Judgments are not being made as to whether they are true or not, but rather that these beliefs affect our decision making.

- "We do not have time to perform RCA."
- "Ensuring patient safety is not synonymous with regulatory compliance."
- "It is politically incorrect for a nurse to question a doctor's decisions."
- "Management does not really want to know what happened."
- "I do not want to participate in RCAs, because it may expose me to a malpractice suit."
- "If the vendor said it, it must be true."
- "Literature searches represent facts; what you read is true!"
- "This is another program of the month."
- "We do not need data to support RCA, because we know the answer. This RCA stuff is a waste of my time."
- "This is another way for administration to 'witch hunt.'"
- "Patients die; that is the nature of healthcare."
- "The hospital is a safe haven in the eyes of the patients, and all their trust is placed in the caregivers' hands."

Many of these statements may sound familiar. But think about how each statement could affect problem-solving abilities. Consider these if–then statements.

- If we see RCA as an added burden (and not a tool), then we will not give it a high priority.
- If we believe that administration values profit more than patient safety, then we may rationalize at some time that bending the rules is really what our administration wants us to do.
- If a nurse fears being reprimanded for questioning a doctor's order, the nurse will not do so when a situation arises in which questions would be appropriate.
- If we believe that participating on an RCA will expose us to legal liabilities, then we will not participate.
- If we believe that RCA is the program of the month, then we will wait it out until the fad goes away.

- If we do not believe that data collection is important, then we will rely on word of mouth and allow ignorance and assumption to penetrate an RCA as fact.
- If we believe that RCA is a witch-hunting tool, then we will not participate.
- If we believe that patients' dying is the nature of this business, we will rationalize this to the point of acceptance into the norm.
- If we believe the hospital is a safe haven, then we will depend on others for our safety.

The purpose of these "if–then" statements is to show the effect that paradigms have on human decision making. When human errors in decision making occur, they are the triggering mechanism for a series of subsequent errors until the undesirable event surfaces and is recognized.

Now the details of the error-change phenomenon and the 5 Ps have been discussed. How do we get all of this information? When an RCA has been commissioned, a group of data collectors must be assembled to brainstorm what data will be necessary to start the analysis. This first session is just that—a brainstorming session of data needs. This is not a session to analyze anything. The group must be focused on data needs and not be distracted by a premature search for solutions. The goal of this first session should not be to collect 100% of the data needed; rather, attempts should be made to be in the 60 to 70% range. All of the obvious surface data should be collected first, and also the most fragile data. Figure 7.5 describes the normal fragility of data at an event scene. The term *fragility* means the prioritization of the 5 Ps in terms of which is most important to collect first, second, third, and so on.

You will notice that "people" and "position" are in a tie position for first item of importance. This is not an accident. As discussed earlier in this chapter, the need to interview observers is immediate in order to obtain direct observation. Positional information is equally important, because it is the most likely to be disturbed the most quickly. Therefore, attempts to get such data should be performed the soonest. Parts are second because if there is not a plan to obtain them, then they will probably end up in the trash. Paper data is generally static, with the exception of process or on-line monitoring systems. Such new technologies allow for automatic averaging of data to the point that, if the information is not retrieved within a certain time frame, it can be lost forever. Paradigms are last, because we wish we could change them faster, but modifying behavior and belief systems takes time.

Analysts should always have a data collection bag prepared. Events usually occur when least expected. The analyst does not want to have to be running around collecting a camera, plastic zip-lock bags, etc. If the tools are all in one place, it is

5Ps	Fragility Ranking
Parts	2
Position	1
Paper	3
People	1
Paradigms	4

FIGURE 7.5   Data fragility rankings.

much easier to go prepared on a moment's notice. Models are available from other emergency response occupations such as doctors' bags, the equipment on fire trucks, the equipment in an ambulance, etc. They always have most of what they need accessible at any time. Such a bag may have the following items for an RCA analyst:

- Digital camera
- Caution tape
- Masking tape
- Zip-lock plastic bags
- Latex gloves
- Scrubs
- Personal protective equipment
- Adhesive labels
- Marking pens
- Video camera (if possible)
- Tweezers
- Pad and pen
- Ruler
- Scalpel for scraping residues
- Sample vials

This is, of course, a partial listing; depending on the organization and nature of work, other items could be added to or deleted from the list.

The following form (see Figure 7.6) is a typical data collection form used for manually organizing data collection strategies for an RCA team.

Data Type: _____ Responsible: _____

Completion Date:_ _ / _ / _

Person to Interview/Data to be Collected: _____
_____
_____

Data Collection Strategy: _____
_____
_____
_____

Task Complete: Yes _____ No _____

Time Taken to Collect Data: _____ Hrs/Minutes

Additional Costs In Collecting Data: $ _____

FIGURE 7.6   Sample 5 Ps data collection form.

1. *Data Type*: List which of the 5 Ps this form is directed at. Each "P" should have its own form.
2. *Responsible*: The person responsible for making sure the data is collected by the assigned date.
3. *Completion Data:* During brainstorming session list all data necessary to collect for each "P" and assign a date by which the data should be collected.
4. *Data Collection Strategy*: This space is for actually listing the plan of how to obtain the previously identified data to collect.
5. *Date to be Collected By*: Date by which the data is to be collected and ready to be reported to the team.
6. *Task Complete:* This space is to acknowledge whether the task has actually been accomplished.
7. *Time Taken to Collect:* It is important to acknowledge how much of people's time was taken to participate on such RCA teams. This number can be applied to a pay rate to determine the actual costs of conducting the RCA and the potential ROIs.
8. *Additional Costs in Collecting Data:* Ancillary costs such as meeting time costs, copying patient records, additional lab costs, the use of external consultants, etc. are all costs associated with conducting the RCA and should be factored into the ROI calculation.

Figure 7.7 is a completed sample data collection worksheet.

Data Type: _Paper_____   Responsible: _John Smith____

Completion Date: 02 /02 /07

Person to Interview/Data to be Collected: _Patient Record____
Shift Logs_____

Data Collection Strategy: _Have nurse supervisor obtain_____
_the patient chart and nurse logs for John Smithson and_____
_deliver to the Risk Manager within 24 hours._____

Task Complete: Yes   X   No ___

Time Taken to Collect Data: _2_ Hrs/Minutes

Additional Costs In Collecting Data: _____
$900 Copying of Patient Record and Nurse Logs_____

FIGURE 7.7   Completed data collection form.

IV

*The PROACT Root Cause
Analysis (RCA) Methodology*

8 Ordering the Analysis Team

Typically, when a sporadic undesirable event occurs in any organization, an immediate effort is organized to form a task team to investigate *why* such an event occurred. What is the typical makeup of such a task team? What should it be?

A patient has had a stay in the hospital extended for six days. This is the result of an allergic reaction that required a medication given by the staff. This is a very simple but routine example.

What happens next? A decision has to be made as to whether this is a reportable incident. A decision will be made as to whether the patient should know (if the patient does not already). If the incident is not reported and the patient is not informed about it, then it may never be known to others or acted on by those involved.

For the sake of argument, let us assume that the allergic reaction to the medication is deemed to be a reportable incident that triggers an RCA to be done. According to policy, the patient will also be informed that the incident occurred and told what will happen as a result. The risk manager (RM) in this hospital is responsible for reporting such events to the appropriate regulatory agencies.

The RM will be the principal analyst (PA) and will assemble a core team. The core team in this case is composed of only two nurses. One of the nurses was the one who administered the medication to the patient. The team brainstorms as to how this could have happened and comes up with the root cause.

Think about what just went on with that team. Remember the earlier discussion about paradigms and how people view the world. How do the nurses likely view the world in this case? They all share the same "box." They have similar educational backgrounds, similar experiences, similar successes, and similar training. That is what they know best: nursing. Anytime a team consists of people with the same occupations and backgrounds, the outcome is predictable. Think about it. If there are five doctors on the team, it will be the nurse's fault. If there are five nurses on a team, it will be the doctor's fault. If doctors and nurses are on the team, it will be administration's fault. The same goes for any expertise in any discipline. This is the danger of not having technical and professional diversity on a team.

To avoid this trap of narrow-minded thinking, the anatomy of an ideal RCA team will be explored. The purpose of a diverse team is to provide synergism where the whole is greater than the sum of the parts. Anyone who has participated in survival teaming games and outings will agree that when different people of diverse backgrounds come together for a common purpose, their outcomes are better as a team than if they had pursued the problem individually.

Teams have long been a part of quality circles and are now commonplace in organizations. Working in a team can be the most difficult part of a work environment, because working with others who may not agree with our views can be difficult. This is why teams work: people disagree. When people disagree, each side must make a case to the other why their perspective is correct. To support this view, a factual basis must be provided rather than conventional wisdom. This is where the learning comes in and teams progress. If a team is moving along in perfect harmony, then changes need to be made in team makeup, as constructive debate is necessary for progress. While this may seem difficult to deal with, it will promote success of the team's charter.

WHAT IS A TEAM?

> A team is a small number of people with complementary skills who are committed to a common purpose, performance goals, and approach for which they hold themselves mutually accountable.*

A team is different from a group. A group can give the appearance of a team; however, the members act individually rather than in unison with others.

Let's now explore the following key elements of an ideal RCA team structure:

Team member roles and responsibilities
Principal analyst characteristics
The challenges of RCA facilitation
Promote listening skills
Team codes of conduct
Team charter
Team critical success factors
Team meeting schedules

TEAM MEMBER ROLES AND RESPONSIBILITIES

Many views about ideal team size exist. The situation that created the team will generally drive how many members are appropriate. However, from an average standpoint for RCA, it has been our experience that between three and five members are ideal, and beyond ten is too many. Having too many people on a team can force the goals to be delayed due to the dragging on of too many opinions.

Who are the core members of an RCA team? They are the principal analyst (PA), the associate analyst, the experts, vendors, and critics.

The Principal Analyst

Each RCA team needs a leader. This is the person who will ultimately be held accountable by administration for results. They are the people who will drive success and accept nothing less. It is their desire that will either make or break the team.

* Katzenbach, Jon R. and Smith, Douglas K., *The Wisdom of Teams,* Harvard Business School Press, Boston, 1994.

   This person is responsible for the administration of the team efforts, the facilitation of the team members according to the PROACT philosophy, and the communication of goals and objectives to senior administration personnel.

The Associate Analyst

This position is often seen as optional; however, if the resources are available to fill it, it is of great value. The associate analyst is basically the "gopher" for the PA. This position is often filled by assistants. This person will execute many of the administrative responsibilities of the PA, such as inputting data, issuing meeting minutes, arranging for conference facilities, arranging for audiovisual equipment, obtaining paper data such as records, etc. This person relieves much of the administration from the PA, allowing the PA more time to focus on team progress.

The Experts

This is basically the core makeup of the team. These are the individuals whom the PA will facilitate. They are the "nuts and bolts" experts on the issue being analyzed. These individuals will be chosen based on their backgrounds in relation to the event being analyzed. If an undesirable outcome in a hospital setting is being analyzed, doctors, nurses, lab personnel, and quality/risk management personnel may serve on the core team. In order to develop accurate hypotheses, experts are absolutely necessary on the team. Experts will aid the team in generating hypotheses and also verifying them in the field.

Vendors

Vendors are an excellent source of information about their products. However, they should not lead such an analysis when their products are involved in the event being analyzed. Under such circumstances, conclusions drawn by the team should be unbiased so that they have credibility. It may be difficult for vendors to be unbiased about how their product performed during use. For this reason, it is suggested that vendors participate on the team but not lead the team.

   Vendors are great sources of information for generating hypotheses about how their products could not perform to expectations. However, they should not be permitted to prove or disprove their own hypotheses. The vendor will usually blame the way in which the product was handled or maintained by the user as the cause of its nonperformance and state that the problem was something that the customer did rather than a flaw in the product. From an unbiased standpoint, all possibilities must be explored, including the possibilities that the product has a problem or that the customer could have done something wrong. Remember, facts lead such analyses, not assumptions!

Critics

Locating critics in any working environment is rarely difficult. Everyone knows who these people are in the organization. However, sometimes the critics get a bad reputation just because they are curious. Critics are typically people who just do not see the world the way that everyone else does. They are really the "devil's advocates" on the RCA teams. They will force the team to see the other side of the tracks and find holes in logic by asking persistent questions. They are often viewed as uncooperative and not team players. But they are a necessity to a team and they promote progress.

Critics come in two forms: (1) constructive and (2) destructive. Constructive critics are essential to success and are naturally inquisitive individuals who take nothing (or very little) at face value. Destructive critics stifle team progress and are more interested in overtime and donuts versus successfully obtaining the team charter.

PRINCIPAL ANALYST CHARACTERISTICS

The principal analyst (PA) typically has a hard row to hoe. If RCA is not part of the culture, then the PA is going against the grain of the organization. This can be very difficult if the analyst has difficulty in dealing with barriers to success. Over the years, we have noted the personality traits that make certain PAs stars, whereas others have not progressed. Below is a list of the key traits that the most successful analysts portray (many of them have led the analyses listed in the case histories of this text).

- *Unbiased*: This is a key trait to the success of any RCA. The leader of an RCA should have nothing to lose and nothing to gain by the outcome of the RCA. This ensures that the outcomes are untainted and credible.
- *Persistent*: Individuals who are successful as PAs are those who do not give up in times of adversity. They do not retreat at the first sign of resistance. When they see roadblocks, they immediately plan to go through them or around them. "No" is not an acceptable answer. "Impossible" is not in their vocabulary. They are painstakingly persistent and tenacious.
- *Responsible* for organizing all the information being collected by the team members and putting it into an acceptable format for presentation. Such skills are extremely helpful in RCA.
- *Diplomatic*: PAs will encounter situations where upper-level management or lower-level individuals will not cooperate. Whether it is unions boycotting teaming, administrations not willing to provide information, or doctors not willing to participate on teams due to legal counsel advice, political situations will arise. A great PA will know how to handle such situations with diplomacy, tact, and candor. The overall objective in all these situations is to get what the team wants. All efforts are to work backward from that point in determining the means to attain the end.

THE CHALLENGES OF RCA FACILITATION

Anyone who has ever facilitated any type of team can appreciate the need to possess the characteristics described above. They can also appreciate the experience that such tasks have provided us about dealing with the human being. Common challenges faced when facilitating a typical RCA team are explored below.

Bypassing the RCA Discipline and Going Straight to Solution

As most people have experienced in their daily routines, the pressure of the daily operation overshadows our intentions of doing things right and stepping back and looking at the big picture. This phenomenon becomes apparent when an RCA team has been organized that is well versed in how to quickly get normal operations up and running. Such teams will be inclined to pressure the RCA facilitator to hurry up and implement his solution(s) as opposed to wasting their time on this "analysis" stuff.

Floundering of Team Members

One of the more predominant problems with most RCA attempts is lack of discipline and direction of RCA methodology being used. This results in the team becoming frustrated, going around in circles and getting nothing done. Also, team members who have not been educated in the RCA methodology being employed can't see the light at the end of the tunnel. Such team members tend to get bored quickly and lose interest.

Acceptance of Opinions as Facts

This often occurs using methodologies that promote solutions before proving that hypotheses are factual. The pressure can be so great to get back to normal that there are tendencies to accept people's opinions as facts so that the team can come to consensus quickly and try to implement solutions. This haste can result in spending money that does not solve the problem, in a "trial-and-error" approach. The true test is, "Would the evidence hold up in court?"

Dominating Team Members

In most teams that are organized under any circumstances there is usually one strong-willed person who tends to impose their personality and authority on the rest of the team members. This can result in the other team members being intimidated and not participating, or it may pressure them to accept opinion as fact, as described above. Doctors have a tendency to fill this role when their subordinates comprise the rest of the RCA team. The doctors sometimes are not even aware that they are doing this.

Reluctant Team Members

Most of us have participated on teams where some members were much more introverted than others. It may not be because they do not have the experience or talent to contribute, but their personalities are simply not outgoing. Sometimes people are reluctant to participate because they feel that authority is in the room; they do not want to appear as not asking the right questions, so they say nothing at all and "do not rock the boat."

Going Off on Tangents

Again, these challenges can (and do) exist on any team. They are functions of team dynamics that happen when humans work together. An RCA facilitator is charged with sticking to the discipline of the RCA method and moving forward, not providing a soapbox for people to hear themselves talk. This includes keeping the team on track and not letting the focus drift.

Arguing Among Team Members

Nothing can be more detrimental to a team than its members engaged in destructive arguments due to closed-mindedness. There is a clear difference between argument and debate. Arguments tend to get polarized, where each side takes a stance and will not budge. The goal of an argument in these cases is for the other side to agree with you totally, not to come to consensus. Debate promotes consensus, which requires a willingness to meet in the middle if necessary.

PROMOTE LISTENING SKILLS

Much of the team dynamics issues discussed thus far are not just pertinent to RCA, but to any team. While the concept of listening seems simplistic, most people are not adept at its use.

Many of us claim that we are not good at remembering names. This is because we don't always listen to people when they introduce themselves. We are often more preoccupied with preparing our response than with actually listening to what the person is saying.

Next time you meet someone, pay attention to the introduction and take a snapshot memory of the person's face with your eyes. You will be amazed at how that impression will log into to your long-term memory and pop up the next time that you meet.

The following are listening techniques that may be helpful when we organize our RCA teams.

One Person Speaks at a Time

This rule is common sense and a matter of simple respect, but how often is this rule broken? It is difficult to listen if there is input from more than one person at a time.

Don't Interrupt

Interruption is rude, and people should be allowed to finish their points while the rest listen. There should be plenty of time to formulate an educated response after the speaker has made his or her point. It may be felt that whoever is the fastest to make statements, or the loudest, will gain ground.

React to Ideas, Not People

This is very important. Even if some team members disagree with others, it should *never* become a personal issue. This is unproductive and will cause digression rather than progression if permitted to exist.

Separate Facts from Conventional Wisdom

Just like in the courtroom, in debates, team members must separate facts from conventional wisdom. The entire discipline in RCA is based on facts. Conventional wisdom originates from opinions and, if not proven, will result in assumptions.

TEAM CODES OF CONDUCT

Codes of conduct are most popular within the quality circles and the push for teaming. They vary from company to company, but what they all have in common is a desire to make meetings more efficient and effective. Codes of conduct are merely sets of guidelines by which a team agrees to operate. Such codes are designed to enhance the productivity of meetings. The following are a few examples of commonsense codes of conduct.

- All members will be on time for scheduled meetings.
- All meetings will have an agenda that is to be followed.
- Everyone's ideas will be heard.
- Only one person speaks at a time.
- "Three-knock" rule—A person politely knocks on the table to provide an audio indicator that the speaker is going off track of the topic being discussed.
- "Holding area"—A place on the easel pad where topics are placed for consideration on the next meeting agenda because they are not appropriate for the current meeting.

This is a sample of some team meeting guidelines. Many organizations will have such codes of conduct framed and posted in all conference rooms. This provides a visual reminder that will encourage people to abide by such guidelines in an effort not to waste valuable time during meetings.

TEAM CHARTER

The team's charter (sometimes referred to as a *mission statement*) is a one-paragraph statement delineating why the team was formed. This statement shall serve as the focal point for the team. Such a statement should be agreed upon not only by the team but also by the managers overseeing the team's activities. This will align everyone's expectations as to direction and results.

The following is a sample charter reflecting a team that was organized to analyze a reportable sentinel event:

> To identify the root causes of the unexpected death of a patient during an inter-hospital transfer, which includes identifying deficiencies in, or lack of, organizational systems. Appropriate recommendations for all identified root causes will be communicated to administration for rapid resolution.

TEAM CRITICAL SUCCESS FACTORS

Critical success factors (CSFs) are guidelines by which success will be measured. CSFs may also be referred to as key performance indicators (KPIs) and other nomenclature, depending on the industry. Regardless of what they are called, there should be some parameters set that will define the success of the RCA team's efforts. This should not turn into an effort in futility by listing 100 different items. It is recommended no more than eight items be designated per analysis. Experience will support that most of this short list of items are used over and over again on various RCA teams. The following are a few examples of CSFs.

- A disciplined RCA approach will be utilized and adhered to.
- A cross-functional section of clinical/nonclinical personnel will participate in the analysis.
- All analysis hypotheses will be verified or disproved with factual data.
- Administration agrees to fairly evaluate the analysis team's findings and recommendations upon completion of the RCA.
- No one will be disciplined for honest mistakes.
- A measurement process will be used to track the progress of implemented recommendations.
- The analysis shall be performed in compliance with all applicable federal, state, and local regulations.

TEAM MEETING SCHEDULES

The question "What is the average time of an RCA?" is often asked. The answer depends on how important the resolution of the event is. The higher the priority of the event being analyzed, the quicker the process will move. Some high-priority events require full-time teams, and resources and funds may be unlimited. These are usually situations where there must be a visual demonstration of commitment on behalf of the organization, for instance because the event has been picked up by the media and now the public wants an answer. These are usually analyses of sporadic events versus chronic. The Challenger and Columbia space shuttle disasters are such examples, where the public's desire to know forced an unrelenting commitment by the government.

This level of attention is rarely given to events that do not hurt individuals, do not destroy equipment, and do not require analysis due to regulatory compliance. These parameters are usually indicative of chronic events.

The intent of this text is to provide the architecture of a sound RCA methodology. It will not work the same in every organization. The model or framework should be molded to each culture to which it is being applied. In essence, we do the best we can with what we have. The process flow involved with RCA team activities might look like Figure 8.1.

It is easy to have idealistic ideas about how RCA teams should function, but in the real world few situations are ideal. As discussed throughout this text, resources and capital are tight while financial expectations rise. This environment does not make a strong case for organizing teams to analyze why things go wrong.

A reality check of how RCA teams will perform under the described conditions is necessary. How will teams realistically deal with analyzing chronic events? Remember that the chronic events are typically viewed as acceptable, they are part

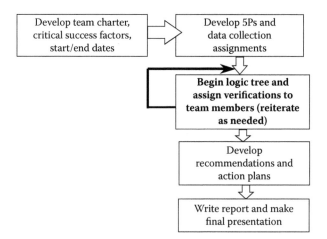

FIGURE 8.1   RCA team process flowchart.

of the budget, and they generally do not hurt people or cause massive amounts of damage to equipment. However, they cost the organization the most in losses on an annual basis.

Assume that opportunity analysis (OA) has been performed, and the "significant few" candidates have been determined. These will likely be chronic versus sporadic events. Now a team has been formed utilizing the principles outlined in this chapter. Where do they go from here?

The first meeting of an RCA team should be to define the structure of the team and delineate the team's focus. As described in this chapter, the team should first meet to develop its charter, critical success factors (CSFs), and the anticipated start and completion dates for the analysis. This session will usually last anywhere from one to two hours. At the conclusion of this meeting, the team should set the next meeting date for as soon as possible.

Because the nature of the events is chronic, the frequency factor is in our favor. From a data collection standpoint, this means opportunity, because the event is likely to occur again. Knowing that the occurrence is likely to happen again, plans can be made to collect data about the event.

This brings us to the purpose of the second meeting of the team: to develop a data collection strategy as described in Chapter 7. Such a brainstorming session should be the second meeting of the team. This meeting typically will take about one to two hours and should be scheduled when convenient for the team members. The result of this meeting will be assignments given for members to collect various types of information by a certain date. At the end of this meeting, the next meeting should be scheduled. The time frame will be dependent on when the information can be realistically collected.

The next meeting will be the first of several involving the delineation of logic utilizing the "logic tree" described in the next chapter. These sessions are reiterative and involve the thinking out loud of "cause-and-effect" or "error-change" relationships. The first meeting of the logic tree development will involve about two hours of developing logic paths. The team should only drill down about three to four levels on the logic tree per meeting. This is typically where the necessary data begins to dwindle, and hypotheses require more data in order to prove or disprove them. The first logic tree-building session will incorporate the data collected from the team's brainstorming session on data collection. The entire meeting usually takes about four hours. About two hours is typically spent on developing the logic tree and another two hours on applying verification information to each hypothesis. At the conclusion of this meeting, a new set of assignments will emerge where verification tests and completion dates will have to be applied to prove or disprove hypotheses. At the conclusion of such logic tree-building sessions, the next meeting date should be set based on the reality of when such test can be completed.

A typical logic tree spans anywhere from 10 to 14 levels of logic. This coincides with the "error-change phenomenon" described in the last chapter. This means that approximately three to four logic tree-building sessions will be required to complete the tree and arrive at the root causes.

To recap, the team will meet on an as-needed basis three to four times for about four hours each to complete the logic tree. It is a myth that such RCA teams are taken out of the field full time for weeks on end. Actual time spent by team members meeting with each other is what is being referred to. This is minimal time relative to the time required in the field to actually collect the assigned data and perform the required tests. Proving and disproving hypotheses in the field is by far the most time-consuming task in such an analysis. However, it is also the most important task if the analysis is to draw accurate conclusions.

By the end of the last logic tree meeting, all the root causes have been identified, and the next meeting date has been set. The next meeting will involve assigning team members to write recommendations or countermeasures for each identified root cause. The teams as a whole will review these recommendations, and they will then strive for a consensus. At the conclusion of this meeting, the last and final meeting date will be set.

The last team meeting will involve writing the report and development of the final presentation. This meeting may require at least one day, because team members are preparing for their "day in court" and they want to have a solid case ready. Typically the principal analyst (PA) will have the chore of writing the report for review by the entire team. The team will work on development of a professional final presentation. Each team member should take a role in the final presentation to show unity in purpose for the team as a whole. The development of the final report and presentation will be discussed at length in Chapter 10.

Each of us has to deal with the reality of our work environments. Keep in mind that the above-described process is just an average for a chronic event. If someone in authority pinpoints any event as a high priority, this process tends to move much faster, as support tends to be offered rather than fought for.

The details of matching the pieces of the puzzle (the data collected) with the ideal team, and making sense in a seemingly chaotic situation, will be explored next.

9 Analyzing the Data
Introducing the Logic Tree

No matter what methods are out in the marketplace to conduct root cause analysis (RCA), they all have one thing in common: cause-and-effect relationships! This is the aspect of science that makes finding root causes possible. The various RCA methods may differ in presentation and expression, but the legitimate ones are merely different in the way in which they derive and graphically represent the cause-and-effect relationships. Everyone will have a favorite "graphical RCA tool," which is fine, as long as people are using it and producing successful results.

In this chapter, the PROACT® RCA tool of choice is called a "logic tree." This is the means of organizing all the data collected thus far and putting it into an understandable and logical format for comprehension and communication. This is different from the traditional logic diagram and a traditional fault tree. A logic diagram is typically a decision flowchart that will pose a question and, depending on the answer, will direct the user to a predetermined path of logic. Logic diagrams are popular in situations where the logic of a system has been laid out to aid in human decision-making. For instance, a 911 operator might use a logic diagram when fielding calls.

A fault tree is traditionally a totally probabilistic tool, which utilizes a graphical tree concept that starts with a hypothetical event. For instance, a person may be interested in how that event *could* occur so he could deduce the possibilities on the lower levels.

A logic tree is a combination of both of the above tools. The answers to certain questions will lead the user to the next-lower level. However, the event and its surrounding modes will be factual versus hypothetical statements. The basic logic tree architecture looks like Figure 9.1. The architecture will be dissected to gain a full understanding of each of its components in order obtain a full comprehension of its power.

BASIC LOGIC TREE ARCHITECTURE

THE EVENT

This is usually the last consequence or effect of an incident, which triggers the analysis to be conducted. This is an important block, because it sets the stage for the remainder of the analysis. *This block must be a fact.* It cannot be an assumption. To determine the event, the analyst should ask the following question: "What consequence triggered this analysis to be conducted?" Under certain conditions, we will accept such an undesirable outcome, whereas in other conditions we will not.

For example, a patient may have been prepped for surgery on the right ankle. However, once in surgery, the surgeon recognizes that the left ankle is marked with the magic marker, indicating it to be the focus of the surgery. The surgeon stops and requests the background data on the patient, which would include all records and

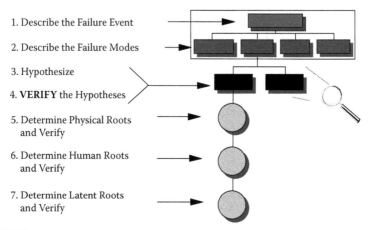

1. Describe the Failure Event

2. Describe the Failure Modes

3. Hypothesize

4. **VERIFY** the Hypotheses

5. Determine Physical Roots
   and Verify

6. Determine Human Roots
   and Verify

7. Determine Latent Roots
   and Verify

FIGURE 9.1   Logic tree architecture.

films from past tests. The surgeon realizes the wrong ankle is marked for surgery. What is the event in this case?

"What consequence triggered this analysis to be conducted?" Because the surgery did not take place and cause undue harm to the patient, it would not necessarily be considered a sentinel event under the TJC guidelines. Therefore, an RCA would not be required from a regulatory standpoint and would not trigger an RCA to be conducted.

However, a progressive facility would recognize this as a near-miss event that easily could have led to a wrong-site surgery that would have resulted in a sentinel event. It would be understood that near-miss investigations would uncover the precursors that could also lead to sentinel events. For this reason, the facility's internal guidelines require an RCA to be conducted on all near-miss incidents. In this case, the trigger would be the near-miss wrong-site surgery instead of an actual sentinel event.

Consider a business example to demonstrate this type of event. When in a business environment that is not sold out (meaning it cannot sell all it can make), there is a tendency to be more tolerant of equipment failures that restrict their ability to produce, because the business cannot sell all it can make anyway. However, when sales pick up and the additional production is needed, it cannot tolerate such stoppages and production rate restrictions. In the non-sold-out state, the event may be acceptable. In the sold-out state, it is not acceptable. This is what was meant by the event being defined as "the reason that triggered the RCA." We only cared here because we could have sold the product for a profit.

In Chapter 7 there was a discussion about how error-change relationships are synonymous with cause-and-effect relationships. The event is essentially the last link in the error chain. It is the last effect or consequence and usually is how we noticed that something was wrong (see Figure 9.2).

Sentinel Events

FIGURE 9.2   Event example.

The Mode(s)

The modes are a further description of how the event *has* occurred in the past. *Remember, the event and mode levels must be facts.* This is the equivalent of asking, "How did the event occur?" It is taking a short step of logic backward in the time line continuum (one link back in the chain of events). This separates the logic tree from a fault tree. It is a deduction from the event block and seeks to break down the bigger picture into smaller, more manageable blocks. Modes are typically easier to delineate when analyzing chronic events. Figure 9.3 is a typical top box (event plus mode level).

When dealing with sporadic events (one-time occurrences), there is no repetition (frequency), so the analyst must rely on the facts at the scene. Assume that a one-time event occurred in which the hospital experienced an extended length of stay due to a patient's allergic reaction to medications administered in the hospital (Figure 9.4).

The Top Box

The top box is the aggregation of the event and the mode levels. As has been emphatically stated, *these levels must be facts!* There is often a propensity to act on assumptions as if they are facts in dealing with RCA teams. This assumption and subsequent action can lead an analysis in a completely wrong direction. The analysis must begin with facts that are verified, and conventional wisdom, ignorance, and opinion should not be accepted as fact based on face value.

The Hypotheses

As we learned in school at an early age, a hypothesis is merely an "educated guess." Without making it any more complex than that, hypotheses are responses to the "how can?" questions described previously. The "how can?" questioning begins when the tree becomes hypothetical, which occurs after the modes (facts) have been established. The answers sought should be as broad and all-inclusive as possible. As will be found in the remainder of this chapter, this is contrary to normal problem-solving tendencies.

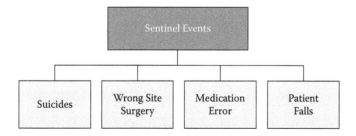

FIGURE 9.3   Top box example for sentinel events.

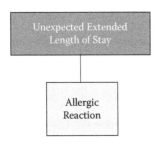

FIGURE 9.4 Top box example for extended length of stay.

Verifications of Hypotheses

Hypotheses that are accepted without validation are merely assumptions. This approach, though a prevalent problem-solving strategy, is really no more trial and error. In other words, it appears to be this case, so we will spend money on this fix and see if it works. When that does not work, the process is reiterated and more money is spent on the next likely cause. This is an exhaustive and expensive approach to problem solving.

In the PROACT methodology, all hypotheses must be validated and supported with hard data. The initial data for this purpose was collected in our 5 Ps effort (parts, position, people, paper, and paradigms). The 5 Ps data will ultimately be used to validate hypotheses on the logic tree. The same approach is used by the police detective preparing for court. The detective seeks a solid case, and so does the analyst. A solid case is built on facts, not assumptions. Would a detective be expected to win a murder case on the basis of a drug dealer's statement alone? This would be a weak case with little likelihood of success.

Hard data for validation means eyewitness accounts, statistics, certified lab tests, surveillance videos, visual inspections, on-line measurement data, and so on. A hypothesis proven with hard data becomes a fact. In keeping with the "solid case" analogy, one must keep in mind that organization is a key to preparing the case. To that end, a verification log should be maintained on a continuing basis to document the supporting data. Table 9.1 provides a sample of a verification log used in RCA. This document supports the logic tree and allows it to stand up (especially in court).

The Fact Line

The fact line starts below the mode level, because above it are facts and below it are hypotheses. As hypotheses are proven and become facts, the fact line moves vertically down the length of the tree. For instance, in the case of the medication error mode described earlier, the fact line might look like Figure 9.5.

TABLE 9.1
Sample Verification Log

Hypothesis	Verification Method	Responsibility	Completion Date	Outcome	Confidence
Prescription Error	Review Prescription	ELZ	04/04/07	FALSE	0
Dispensing Error	Review Prescription, Order Entered and Dispensed	TGJ	04/06/07	TRUE	5
Monitoring Error	Review Strips and Logs	RPG	04/01/07	FALSE	0
Medication Stockout	Review Med Supply Logs	RCA	04/02/07	FALSE	0

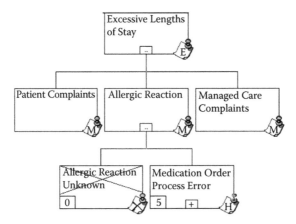

FIGURE 9.5   The fact line positioning.

PHYSICAL ROOT CAUSES

The first root level causes encountered through the reiterative process will be the physical roots. Physical roots are the tangible roots or component-level roots. When undisciplined problem-solving methods are used, people will have a tendency to stop at this level and call them "root causes." All physical root causes must endure validation to prove them as facts. Physical root causes are typically the first physical consequences of improper actions taken by the human being.

HUMAN ROOT CAUSES

Human root causes will almost always trigger a physical root cause to occur. Human root causes are decision errors that result in errors of omission or commission. This means that one either decided *not* to do something he should have done, or he *did* something he was not supposed to do. Examples of errors of omission might be delay in ED treatment, delay in administering medications on time, or failure to check on a patient due to being distracted by other priorities. An error of commission might be administration of the wrong type, frequency, and/or dose of medication to a patient, an ED doctor misdiagnoses a patient, or surgery is performed on the wrong side of the body.

While the questioning process thus far has been consistent with asking "how can?" at the human root level, the questioning is switched to "why?" When dealing in the physical and process areas, getting answers to such "why?" questions is not possible, as they are inanimate objects and not capable of making decisions. Only at the human root level do we encounter a person. When this level is attained, the analyst is not interested in who made the decision but rather why they made the decision they did. Understanding the rationale behind decisions that result in error is the key to conducting true RCA. Anyone who stops an RCA at the human level and disciplines a person or group without basis is participating in a "witch hunt." Witch hunts were discussed in Chapter 7 and proven to be non-value added, as the true root cannot be attained in this manner. If the analyst is searching for a scapegoat, no one will want to participate in the analysis for fear of repercussions. An analyst who cannot find out why people make decisions cannot solve the issue at hand and prevent it from recurring.

LATENT ROOT CAUSES

Latent root causes are the organizational systems used to make decisions. When these systems are flawed, they result in decision errors. The term "latent" * is defined as:

> ...whose adverse consequences may lie dormant within the system for a long time, only becoming evident when they combine with other factors to breach the system's defenses.

When the term *organizational* or *management* systems is used, it is referring to the "rules and laws" that govern a facility. Examples of organizational systems might include policies, operating procedures, medical equipment management procedures, maintenance procedures, purchasing practices, stores and inventory practices, etc. These systems are all in place to help people make better decisions. When a system is inadequate or obsolete, people end up making decision errors based on flawed information. These are the true root causes of undesirable events.

The most relevant terms have now been defined regarding the construction of a logic tree. Now the physical building of the tree and the thought processes that go on in the human mind will be explored.

Experts who participate on such RCA teams are generally well-educated and well respected within the organization. Using the logic tree format, an expert's thought process may look like Figure 9.6.

The tendency to jump from problem to cause, without analysis, poses another hurdle for the RCA team. It is the responsibility of the principal analyst (PA) to manage the expertise of the team in a constructive manner without alienating the team members. An RCA team will have a tendency to go straight to the micro, rather than the macro, view. However, in order to understand exactly what is happening, the team must step back and look at the big picture. To do this, the team's original thought process must be recreated so the search for assumptions in the logic can begin.

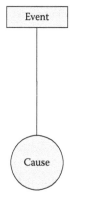

FIGURE 9.6 The "expert's" logic tree.

---

* Reason, James, *Human Error,* Cambridge University Press, London, 1990–1992, 173.

A logic tree is merely a graphical expression of what a thought would look like if it were on paper. It is actually looking at how one thinks. For example's sake, the allergic reaction case discussed earlier will be expanded on (see Figure 9.7).

BROAD AND ALL-INCLUSIVE ANALYSIS

Assume an RCA team of nurses, physicians, and pharmacists was assembled and asked "How could the patient have had an allergic reaction?" Their answers would likely get into such details as: it was a prescription error by the physician, it was a dispensing error by the pharmacy, the nurse administered the wrong medication, etc. While these are all valid points, they jump to too much detail too quickly. The analyst wants to use deductive logic in short leaps.

In order to be "broad and all inclusive" at each level, the analyst must try to identify all the possible hypotheses in the fewest amount of blocks.

In answering the question of how the patient could have an allergic reaction, there are only two practical ways in which a patient could have an allergic reaction: (1) the staff was not aware of the allergic condition of the patient, or (2) the staff knew about the allergic condition of the patient and administered the wrong frequency, type, or dose of the medication. All of the hypotheses developed earlier by the team experts would cause one of these two conditions to exist (see Figure 9.8).

From this point, the team would review some of their 5 Ps data previously collected. A review of the patient record demonstrates that the allergic condition was revealed and acknowledged. The patient record becomes the data source (paper)

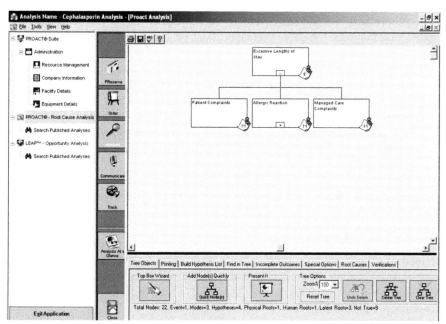

FIGURE 9.7 Extended length of stay "top box."

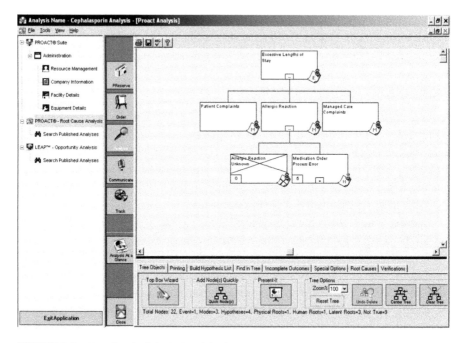

FIGURE 9.8   Broad and all-inclusive thinking.

validating that the patient's allergic condition was known. Thus, the hypothesis "allergic condition unknown" cannot be true.

Consider these two alternatives and the different paths they would have taken the analyst based on what the data concluded.

1. If the patient record revealed no evidence of a known allergic reaction, then the next question in the logic tree would have been, "How could the patient have had an allergic condition and the staff not know about it?"
2. Since the record did reveal that the allergic condition was known, then the next question in the logic tree would be, "How could we have known the patient had an allergic condition and still provided a medication that would trigger it?"

Pursuing the two paths would take the analyst in two different directions. The data leads us in the correct direction, not the analyst. As stated previously, this process is entirely data driven.

If the logic tree is broad and all-inclusive at each level, and each hypothesis is verified with hard data, then the fact line drops to the next level until all root causes have been discovered. This is similar to the quality initiatives of the recent logic tree. The logic tree is ensuring "quality of the process" so, by the time the root causes are determined, they are correct.

THE ERROR-CHANGE PHENOMENON APPLIED TO THE LOGIC TREE

How does the error-change concept (cause-and-effect relationship) parallel the logic tree? As the path of the logic tree is explored, there are three key signs of hope that we can solve the failure. These keys are as follows:

Order
Determinism
Discoverability

ORDER

If one truly believes that the error-change phenomenon exists, then one has the hope that following cause-and-effect relationships backward will lead to the culprits: the root causes. Our students are often asked, "Do you believe there is order in everything, including nature?" There is generally a silent pause while they think about it, and they cite facts such as tides coming in and going out at predetermined times, the sun rising and setting at predetermined times, and the seasons that various geographic regions experience on a cyclical basis. These are all indications that such order, or patterns, exist in nature.

DETERMINISM

This means everything is determinable or predictable within a range. If an IV pump has failed, the reasons (hypotheses) of how the IV pump can fail are determinable. These possibilities are determinable within a range.

People are the same way to a degree. People are determinable within a broader range than equipment because of the variability of the human race. If we subject humans to specific stimuli, they will react within a certain range of behaviors. If employees are alienated publicly, then chances are they will withdraw their ability to add value to their work. They in essence become human robots because of the way they were treated.

Determinism is important because, when constructing the logic treatise, it is essential to develop hypotheses from level to level based on determinism.

DISCOVERABILITY

This is the concept that when you answer a question, it begets another question. The analogy of children in the ranges of 3 to 5 years old demonstrates true discoverability. They make beautiful analysts because of their innocent inquisitiveness and openness to new information. Many of us have experienced our children at this age when they say, "Daddy, why does this happen?" The series of "why" questions can be generally answered about five times (remember the 5 Whys discussion) before we do not know the actual answer. This is discoverability; questions only lead to more questions. On the logic tree, discoverability is expressed from vertical level to vertical level when the analyst asks, "How can [something] occur?" The answer only leads to another "how can?" question.

All of these keys provide the analyst with the hope that there is a light at the end of the tunnel. The analyst is searching for a pattern in a sea of chaos and the above keys help him to find this pattern. Imagine if we were the investigators at the original bombing site of the Twin Towers building in New York in 1993, could we even visualize finding the answer by looking at the enormous amount of rubble generated from blast—the chaos? Yet the investigators knew there was a pattern somewhere in the chaos, and they went about finding it. Within about two weeks of the blast, investigators knew the type of vehicle, the rental truck agency, and the makeup of the bomb. This demonstrates true faith in finding a pattern in chaos. These people believe in the logic of failure (see Figure 9.9).

AN ACADEMIC EXAMPLE

The logic tree architecture described above will now be put into perspective in an academic example that is easy to relate to.

Most people have experienced problems at some time or another with their local area networks (LANs). This is an almost universal issue that happens every day in our high-tech world. Looking at the issue from an RCA perspective, "What is the event in this case?" the end of the error chain is that the LAN is not functioning as it was designed. Therefore, the analyst may say that "recurring LAN failures" is the event, because it is the consequence that trigger us to take action—the last effect or consequence of the cause-and-effect chain (see Figure 9.10).

Moving on to the second level, the modes must be described that are the symptoms of how the analyst knows that the LAN has not been performing as designed. This information, in this situation, may come from users at their workstations in the form of complaints to the information systems (IS) department. Some examples at this level might be slow database access time, hard disk failure, printer fails to print, and no network connection. These are all facts, because the users have observed them in the past.

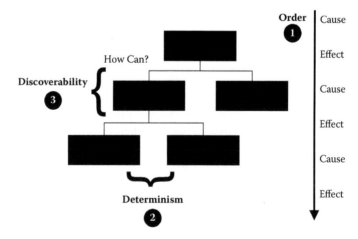

FIGURE 9.9   Three keys.

Which mode would be approached first? Had an opportunity analysis (OA) been performed, such priorities would have already been quantifiably established. In this example, the mode with the greatest frequency of occurrence

FIGURE 9.10 The LAN event block.

will be pursued, and that might be "printer fails to print." In this particular office, the majority of the complaints have been that the printer does not print when users send a job to it. These complaints absorb about 80% of the IS technician's time. For this reason, this leg will be pursued. The *top box* may look like Figure 9.11.

At this point, hypothetical questioning begins with, "How could the printer fail to print?" The natural response would be such answers as no toner, no power, wrong configuration, operator error, no paper, etc. All of these hypotheses are valid, but do they meet the criteria of "broad and all-inclusive?" This is the most difficult portion of constructing a logic tree, thinking broadly! Is it possible that all of the hypotheses would have to be embedded somewhere in the operator, the printer, the computer, or the cable? These four hypotheses are broad and all inclusive. The next level of this logic tree would look something like Figure 9.12.

Now comes the task of proving which hypotheses are true and which are not. It is at this point that the verification log begins to be developed and the information collected in our 5 Ps is utilized as validation data. The analyst will take the first hypothesis of "the printer" and determine a test that can prove or disprove it. The test

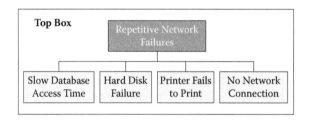

FIGURE 9.11 LAN example top box.

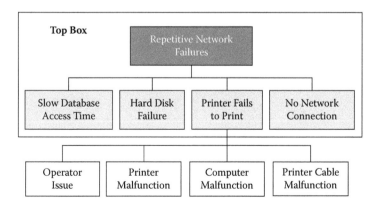

FIGURE 9.12 The first hypothetical leg.

could be that a laptop computer is connected to the printer that is not printing, using the same cable and the same operator to test its functionality. In this case, the printer functions as designed. Based on this test, the hypotheses of "the printer," "the cable," and "the operator" can be crossed out.

However, "the computer" cannot be selected by process of elimination. "The computer" must also have a test to validate it. In this case, another known working printer can be connected to the same computer to test its functionality with the same operator. It is concluded from this test that the new printer also does not perform with the same computer. Based on these tests, the logic tree would look like Figure 9.13.

At this point the fact line has moved down from the mode level to the first hypothesis level. Because the hypothesis of the computer has been verified as "true," it is now a fact, and the fact line drops. Likewise, the other three hypotheses have proven to be false and do not require to be drilled down further.

The questioning continues to the next level by asking, "How could a computer malfunction, causing the printer not to print?" This is the discovery portion of the logic tree where one question begets another. Again, 100 reasons could be brought up as to how a computer could malfunction, but we need to think broadly. The broadest possibilities that meet these criteria may be "hardware malfunction" and/ or "software malfunction." Now tests must be developed to prove or disprove these hypotheses. Running diagnostic software determines that the system is not recognizing a parallel port card. Other than the identified hardware malfunction, there are no indications of any software malfunctions. This allows us to cross out "software malfunction" and continue to pursue "hardware malfunction" (see Figure 9.14).

The reiterative questioning continues with, "How can we have a hardware malfunction that would create a computer malfunction that would not allow the printer to print?" Notice in this questioning that the logic path is always being read back a few levels to maintain the "storytelling" or consequence path. This helps the team follow the logic tree and put the question into proper perspective. The broad and all-inclusive answer here could be either a "system board failure" and/or a "parallel port card malfunction." The previous test of running the diagnostic software confirms an issue with the parallel port card. The motherboard has displayed no signs of malfunction within the context of the entire system. If a problem were apparent with the

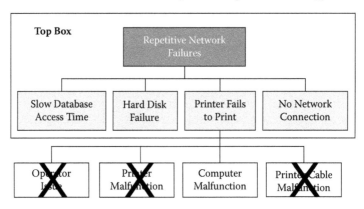

FIGURE 9.13   Updated logic tree.

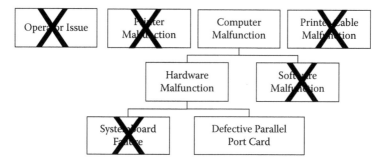

FIGURE 9.14 Hypothesis validation.

motherboard, there would be more apparent issues other than just a printer failing to print. The absence of these issues is the validation that the motherboard does not appear to be a contributor to this event.

While parallel ports used in the above fashion are an outdated technology, this is still an effective example from a cause-and-effect standpoint. At this point, the parallel port card is removed and the contact areas cleaned and reseated back into the appropriate slots, making sure the contact is made and that improper installation concerns are not an issue. The printer still fails to print even when the parallel port card has been confirmed to be installed correctly. Next the parallel port card is replaced with one known to be working and properly installed into the computer. This time the printer works as desired. It may be felt that the analysis is complete at this level, because the event will not occur immediately. However, this is a point where a physical root would be considered, when the event temporarily goes away. This identified block would be labeled as a "defective parallel port card," indicating it as a physical root cause (Figure 9.15).

Having identified the physical root in this case means there is more work to do in order to get down to the latent roots. The questioning continues with, "How can we have a defective parallel port card, which is causing a hardware malfunction, which is causing a computer malfunction, which is causing the printer not to print?"

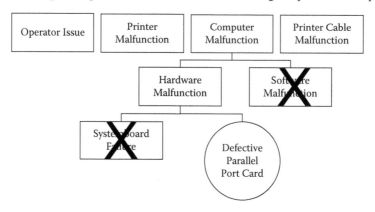

FIGURE 9.15 Identification of physical root cause(s).

It was "installed improperly," and/or it was "purchased in a defective state," and/or it "failed during our use."

It has already been determined that the installation practice was proper. The "card failed during our use" can be eliminated because interviews reveal that this was a new printer added to the network, and it never worked from the beginning. People on the network therefore chose to divert to another network printer. So this was not a case where the printer worked at one time and then did not work. This serves as the proof that the parallel port card did not become defective in our possession, but rather that it was received that way from the manufacturer or vendor.

Now the purchasing practices must be reviewed to determine whether there are purchasing procedural flaws allowing defective parts to enter the organization. From the 5 Ps information, it was determined that there is "no list of qualified vendors," and there are "inadequate component specifications." The primary concern is found to be that purchasing buys based on low cost because that is where the incentives are placed for the purchasing agents in this firm. It has now been confirmed that the organization "purchased a poor quality card" and, because this task involves a conscious human decision that results in an action to be taken, this is deemed to be one of the human root causes. This block is now labeled as a human root. Remember, at this human root level, the questioning protocol switches to "why?" because a human being can now respond to the question. The question at this point would be, "Why did we choose to purchase this particular parallel port card from this vendor at that time?" The potential answers are "no list of qualified vendors," "no component specifications," and "low-cost mentality." These are the reasons behind the decision and therefore are the latent roots (Figure 9.16).

Could this example, although academic, relate to situations in real environments where disruptions are caused in processes due to the infiltration of defective parts and/or supplies into the organization? Without a structured RCA approach, trial-and-error would be utilized until something worked. This is a very expensive option.

What if the analyst stopped at the physical root of "defective parallel port card" and just replaced the card? Would the event likely recur? It would, if the same purchasing habits continued. It may not happen in the same location, because not all cards would be defective, but it would likely happen somewhere else in the organization, forcing another need to solve the same problem.

What if the analyst stopped at the human root of "purchased a defective parallel port card" and disciplined the purchasing agent who made the decision? Would that prevent recurrence? Not likely, because the decision-making system the agent used is probably being used by other purchasing agents in the organization. It might prevent that agent from making such a decision in the future, but it would not stop other such decisions from being made in the future by other agents.

The only way to prevent recurrence of this event throughout the organization is to correct the decision-making systems referred to as the latent roots or organizational system deficiencies. When such deficiencies are uncovered, "root" cause analysis is truly being performed.

The completed verification log for the above example might look like Table 9.2.

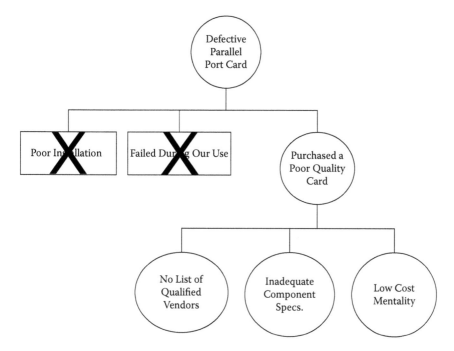

FIGURE 9.16　Identification of human and latent roots.

HYPOTHESIS CONFIDENCE FACTORS

Experience supports that the more data collected initially, the quicker the completion of the analysis and the more the accurate result. Conversely, the less data the analyst has initially, the longer the analysis takes and the greater the risk of the wrong cause(s) being identified.

PROACT utilizes a confidence factor rating for each hypothesis to evaluate the degree of confidence the analyst has with the validity of the verification test and the accuracy of the conclusion. The scale is basic and runs from "0" to "5." A "0" means that with 100% certainty, with the data collected, the hypothesis is *not* true.

On the flip side, a "5" means that, with the data collected and the tests performed, there is 100% certainty the hypothesis is true. Between the "0" and the "5" are the shades of gray where the data used was not absolutely conclusive. This is not uncommon in a situation where an RCA is commissioned weeks after the event occurred and little or no data from the scene was collected. The confidence factor rating communicates this level of certainty and can guide corrective action decisions.

Analysts will have to develop their own rules of thumb, but a common rule is that a confidence rating of "3" or higher is treated as if the hypothesis did happen and it is necessary to continue to pursue the associated logic leg. Any confidence rating of less than "3" is treated as a low probability of occurrence and should not be pursued at this time. However, the only hypotheses that are crossed out are the ones that have a confidence rating of "0." A "1" cannot be crossed out, because it still has a probability of occurring even if the probability is low.

TABLE 9.2
Completed Sample Verification Log

Hypothesis	Verification Method	Responsibility	Completion Date	Outcome	Confidence
Operator Issue	Use same operator to test with new printer	TDF	4/4/07	Printer worked fine on alternate computer using same operator	0
Printer Malfunction	We utilized a stand-alone laptop to test the printer	RMB	4/4/07	Printer worked fine on alternate computer	0
Computer Malfunction	Utilize a known working printer and test in on the computer	TDF	4/4/07	Still could not print after the test	5
Printer Cable Malfunction	We utilized a stand-alone laptop to test the cable	RMB	4/4/07	Cable worked fine on alternate computer	0
Hardware Malfunction	Run diagnostic software to check hardware	TDF	4/5/07	Determined a possible problem with parallel port card	5
Software Malfunction	Check drivers and configuration	TDF	4/6/07	Configuration and drivers were correct	0
Systemboard Failure	Call in a technician to test the system board for faults	TDF	4/7/07	Not indication of system board failure	0
Defective Parallel Port Card	Replace the card with a known working card	RMB	4/8/07	The document printed fine using the alternate card	5
Poor Installation	Check installation notes as well as talking with technician who installed the card	RMB	4/3/07	Installation looked adequate	0
Purchased a Poor Quality Card	Talk with the purchasing department and storeroom personnel	RMB	4/6/07	Determined that this was a new installation and discovered the card never worked properly	5
Failed While in Our Use	Review failure history	RMB	4/5/07	Records indicate that card never worked properly	0
No List of Qualified Vendors	Determine current vendor requirements	JCF	4/6/07	Records determined that we have no list of qualified vendors	5
Low Cost Mentality	Look for a history of low bidder mentality	FRD	4/6/07	This has been the prevalent purchasing practice in the purchasing department	5
Inadequate Component Specifications	Check the component specifications for this and other related items	FGH	4/6/07	We did not have solid component specifications	5

ORGANIZATIONAL MEMORY: THE LOGIC TREE AS A TROUBLESHOOTING FLOW DIAGRAM

Once the logic tree is completed, it should serve as a troubleshooting flow diagram for the rest of the organization. Chances are the causes identified in this RCA will affect the rest of the organization. Therefore, some recommendations will be implemented sitewide. The goal of a world-class RCA effort should be the development of a dynamic troubleshooting flow diagram repository. These flow diagrams will end up being a logic tree that contains the expertise of the organization's best problem solvers in a single database.

Such logic trees can be stored on the company's network and made available to all facilities that have similar processes and can learn from the work done at one site. These logic trees are complete with verification tests for each hypothesis. It is dynamic because, where this RCA team may not have followed one particular hypothesis (because it was not true in their case), it may be true in another case, and the new RCA team can pick up from that point and explore the new logic path.

The goal of the organization should be to capture the intellectual capital of the workforce and make it available for the entire organization. This optimizes the intellectual capital of the organization through RCA and is sometimes referred to as storing "corporate memory."

THE LOGIC TREE APPLIED TO CRIMINOLOGY APPLICATIONS

Since the detective analogy has been used throughout this process, it may be appropriate to highlight the differences in the application of the logic tree to this field.

To quickly refresh our memories to this point: efforts have been made to "preserve" the event data using the 5 Ps. The appropriate team was assembled in the "ordering" the analysis team section, and now the logic tree process was used to graphically depict the cause-and-effect relationships leading to the event being analyzed. How does this correspond to the role of the detective (see Figure 9.17)?

What are the similarities between how a logic tree would be used by a detective and a risk manager? The top box is composed of the event and the modes. These have been described as *factual* information that defines the event. A fact is something that is proven and not left to supposition. What do the detectives have on which to base their case? Their facts generally start with the evidence at the scene of the crime. As discussed earlier, the line around the top box could be synonymous with the yellow police tape around the crime scene. The police tape seals off the area to preserve the facts. The facts ultimately serve as evidence in building the case. The event may be the consequence, which forces police to investigate. Such events might be a murder, a

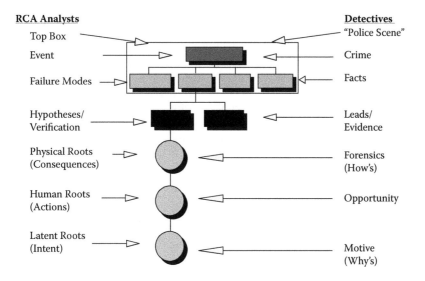

FIGURE 9.17   Logic tree comparison.

robbery, or a suspicious fire. When looking at these events, the detective would have to look at the evidence to determine the nature of the crime. How did the murder occur? Was it by gun, knife, strike by a car, poisoning with medication, etc.? These would represent the modes.

From the facts, leads are generated. When leads are generated in these situations, it is typically because the question was asked, "How could this type of trauma to the head have occurred?" This *leads* them to their hypotheses. Once the hypotheses have been developed, the detectives must now use the skills of forensic scientists to prove or disprove them scientifically. This is akin to healthcare using medical forensic personnel (i.e., medical examiner) to test for chemicals found in body tissues of a patient. Science in both cases is used to determine the *how* of the crime or the event.

When exploring the human and latent root causes, the analyst now ventures off into areas that are not typically the expertise of the forensic scientists. This is because the analyst is now looking for the answer to, "*Why* did this person make the decision to commit a crime?" In police terms, the analyst is now seeking *motive* and *opportunity*. Typically, forensic professionals are involved in the *hows* of an event. The responsibility of determining the *motive* and *opportunity* lies with the investigator and the prosecutor.

The moral of this story is that the *top box* is equivalent to the crime scene. The development of hypotheses and their verification is the equivalent to the *how* in the crime lab. The human and latent root causes are equivalent to establishing *motive* and *opportunity*, or *why* someone made a particular decision.

This logical process is applicable in any situation where humans are involved. The proceses of logical deduction and reasoning are elements of the human decision-making process and the brain's built-in ability to resolve problems. The *nature* of the problem is irrelevant; the process is the same.

THE ROOT SYSTEM CONTRASTED TO RESEARCH

Noted author and researcher in the field of human error James Reason has coined and redefined many terms in this field. In Figure 9.18, Reason uses the terms *intent, actions,* and *consequences* to describe the anatomy of a failure.

These terms (*intent, actions,* and *consequences*) are synonymous with latent, human, and physical root causes. People usually make decisions with good intentions; however, the organizational system they are referencing is flawed in some fashion. Based on this flawed data, the person makes a decision that results in an action or inaction. As a result of this decision, a series of consequences will begin. This series of consequences is the error chain referred to in Chapter 7. This series of consequences continues until the organization has no choice but to respond (event).

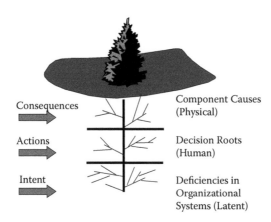

FIGURE 9.18 Human error related to logic tree.

10 Communicate Findings and Recommendations

THE RECOMMENDATION ACCEPTANCE CRITERIA

Assume at this point that the complete RCA process has been followed to the letter. The opportunity analysis (OA) has been conducted and has determined the "significant few." A specific significant event has been selected and proceeded through the PROACT® RCA process. An RCA team has undertaken an organized data collection strategy. The team's charter and critical success factors (CSFs) have been determined, and a principal analyst (PA) has been named. A logic tree has been developed where all hypotheses have been either proven or disproved with hard data. Physical, human, and latent roots have been identified. Is the RCA done?

Not quite! Success can be defined in many ways, but an RCA should not be deemed successful unless something has improved as a result of implementing recommendations from the RCA. Merely conducting an excellent RCA does not produce results. As many can attest, getting something done about RCA findings can be the most difficult part of the analysis. Recommendations will often fall on deaf ears, rendering the entire effort a waste of time and the organization's money.`

If it is known that such hurdles will be evident, they can be planned for proactively. To that end, it is suggested that a "recommendation acceptance criteria" be developed. Most people have faced situations where hours, and sometimes weeks and months, are spent developing recommendations as a result of various projects only to have the recommendations turned down flat by the reviewers. Sometimes explanations are given, and sometimes they are not. Regardless, it is a frustrating experience, and it does not encourage creativity in making recommendations. As a result of this experience, analysts may tend to become more conservative in their recommendations, keeping them to a minimum to merely get by.

A recommendation acceptance criterion is called "rules of the game." Executives are typically charged with fiscal responsibility in any organization. They make the economic decisions as to how the organization's money is spent. Whether these rules are written or unwritten, they define whether or not the RCA recommendations will fly with management. The analyst should ask the approving executives/managers for "the rules of the game" before they begin to come up with recommendations. This is a reasonable request seeking only to not waste valuable and scarce time and money on non-value added work.

A sample listing of recommendation acceptance criteria might look like this:

The recommendation must:

1. *Eliminate or Reduce the Impact of the Cause.* The goal of an RCA may not always be to eliminate a cause.

   Example: If it is found that blood redraws are occurring 10,000 times per year, a realistic expectation may be to initially reduce the number of redraws by 50% in the first year (reducing the impact).

2. *Provide a ___% Return on Investment (ROI).* Most organizations will have either written or unwritten rules dictating the ROI expected for the expenditure of funds, especially capital funds. Ten years ago, such ROIs in industry were frequently around 15 to 20%. More recently, the average range for these numbers and expectations is 50 to 100%. This indicates a risk-aversive culture where only certainty is accepted. Such a culture will limit significant progress in the long term.

   Example: A recommendation to hire a phlebotomy team to take over the blood drawing tasks from the nurses will only be considered if a cost-benefit analysis can demonstrate a 100% ROI in the first year to obtain and sustain such a team.

3. *Not Conflict with Capital Projects Already Scheduled.* Sometimes lengthy recommendations are developed only to find that some strategic plans are on the books, unbeknownst to RCA team. These plans may call for the abandonment of old units or addition of new units in the hospital. If the RCA team is aware of such secret plans, then they will not spin their wheels developing recommendations that have no chance of succeeding.

   Example: If the RCA team is not aware of executive efforts to hire a phlebotomy team to take over the blood drawing tasks, then the team could spend its time developing a recommendation on training current nursing staff in proper blood drawing techniques. This would be a waste of time under this scenario.

4. *List All the Resources and Cost Justifications.* Executives like to know that much thought has gone into how to execute presented recommendations. Therefore, include in the cost/benefit analysis manpower resources required, materials necessary, safety and quality considerations, etc.

   Example: If the recommendation is to hire a phlebotomy team to take over the blood drawing tasks, then the lifecycle costs of such a decision will have to be compared to the current situation. What will be the potential long-term impact of additional salaries, benefits, liabilities, additional exposure, cost of living allowances (COLAs), etc.? The lifecycle costs of the recommendation should justify acceptance of the recommendation when compared to the lifecycle costs of current operations.

5. *Have a Synergistic Effect on the Entire System/Process.* Often, in working environments, "kingdoms" develop internally and stifle communications, and they end up competing against each other for time and resources. This scenario is commonplace and counterproductive. Executives should expect recommendations that are synergistic for the entire organization.

Recommendations should not be accepted if they make one area look good at the sacrifice of other areas.

Example: The hiring of a phlebotomy team may look fiscally responsible, but will it be culturally responsible? Will such a decision demoralize the nursing staff, indicating that executive management did not think them competent to draw blood from patients? Will this demoralization cause experienced nurses to go to competing hospitals? If so, did the decision to hire a phlebotomy team have a synergistic effort on the organization?

Analysts do not want to waste their time and energy developing recommendations that, when presented to the decision makers, do not have a chance of flying. Efforts should be made to seek out such information and the team's recommendations should be framed around the criteria provided.

DEVELOPING THE RECOMMENDATIONS

Every corporation has its own standards regarding how recommendations should be written. It will be the RCA team's goal to abide by these internal standards while accomplishing the objectives of the RCA's team charter.

The core team members, at a predetermined team meeting time and location, should discuss recommendations. The entire meeting should be set aside to concentrate on recommendations alone. At this meeting, the team should consider the recommendation acceptance criteria (if any were obtained) and any extenuating circumstances. Remember our analogy of the detective throughout this text, always trying to build a solid case with the prosecutor. This report and its recommendations represent the team's "day in court." In order to win the case, the recommendations must be solid and well thought out. But foremost, they must be accepted, implemented, and effective in order to be successful.

At this team meeting, the objective should be to gain team consensus on recommendations brought to the table. Team consensus is *not* team agreement. Team agreement means that everyone gets what he or she wants. Team consensus means that everyone can live with the recommendations decided upon. Everyone may not have gotten everything they wanted, but they can live with it. Team agreement is rare.

The recommendations should be clear, concise, and understandable. When writing the recommendations, the objective should be to eliminate or greatly reduce the impact of the cause. Every effort should be taken to focus on the RCA. Sometimes there is a tendency to have pet projects attached to an RCA recommendation, because they might have a better chance of being accepted. This can be likened to riders (pork) on bills being reviewed by Congress. This can bog down a good bill and threaten its passage in the long run. The first sight of a smokescreen by executives will affect the credibility of the entire RCA. When writing recommendations, stick to the issues at hand and focus on eliminating the risk of recurrence.

It is good practice to have a backup alternative to every recommendation that might be questionable. Sometimes when recommendations are developed, they might be perceived as "on the edge" of the criteria given by the executives. If this is the case, then efforts should be taken to have a backup recommendation—a recommendation that clearly fits within the defined criteria. We want to ensure that our final meeting is not held up by something within our control. Not having acceptable recommendations in our meeting is something within our control and should not happen.

DEVELOPING THE REPORT

The report represents the documentation of the "solid case" for court or, in the analyst's circumstances, the final meeting with the executives. This should serve as a living document in that its greatest benefit will be that others learn from it so as to avoid recurrence of similar events at other facilities within the company or organization. To this end, the professionalism of the report should suit the nature of the event being analyzed. If the event costs the corporation $5, then perform a $5 RCA. If it costs the organization $1,000,000, then perform a $1,000,000 type of RCA.

The RCA team should keep in mind that, if true RCAs are not prevalent in an organization, the first RCA report usually sets the standard. The team should take this into consideration when developing its reports. Assume at this point that the team has analyzed a "significant few" type of event and it is costly to the organization, so the report will reflect that level or degree of importance. The following is the team's guide for the report.

EXECUTIVE SUMMARY

The executive summary is just that: a summary for executives to review. The typical decision makers at the executive level are not as concerned with the details of the RCA as with the results and credibility of the RCA. This section should serve as a synopsis of the entire RCA—a quick overview. It a review of the event analyzed, the reason it occurred, what the team recommends to make sure it never happens again, and how much it will cost.

Event Summary

The event summary is a description of what was observed from the point in time the event occurred until the point in time the event was isolated or contained. It is a time line description.

Analysis Findings

The analysis findings present a description of the findings of the RCA. This is a summary of the cause-and-effect relationships that led up to the point in time of the event occurrence. This is meant to give executives a quick understanding of the chain of events that were found to have caused the event in question.

RCA Methodology Description

This is a basic description of the RCA methodology used so the executives can quickly understand the process. Sometimes management may not be aware of a formalized RCA process being used at the sharp end. A basic description of this disciplined and formal process adds credibility to the analysis and assures management that it was not a "buckshot" analysis.

Root Cause Action Matrix

The root cause action matrix is a table outlining the results of the entire analysis. It is a summation of identified root causes, overview of proposed recommendations, people responsible for executing recommendations, and estimated completion dates. Figure 10.1 is a sample root cause action matrix.

Estimated ROI of Recommendations and Analysis

This section would outline briefly the estimated return on investment (ROI) for each proposed recommendation and also for conducting the RCA as a whole.

TECHNICAL ANALYSIS

This is where the details of all recommendations are located. It is where the technical/clinical staff can review the "nuts and bolts" of the analysis recommendations.

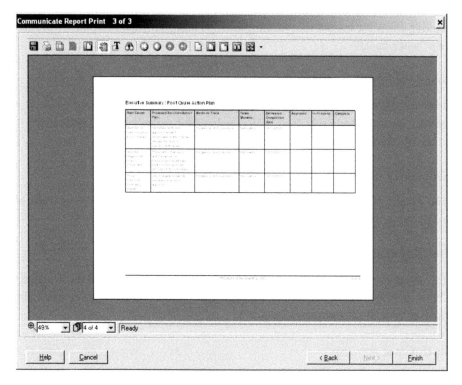

FIGURE 10.1   Sample root cause action matrix.

Identified Root Causes

The identified root causes will be delineated in this section as separate line items. All causes identified in the RCA that require countermeasures will be listed here.

Types of Root Causes

The types of root causes will be listed here to indicate their nature as being physical, human, or latent root causes. Only in cases with "intent with malice" should any indications be made as to identifying any individual or group. Even in such cases, it may not be prudent to specifically identify a person or group in the report because of liability concerns. Normally, no recommendations are required or necessary where a human root is identified. This is because if the latent root or the decision-making basis that led to the occurrence of the event was addressed, then subsequently the behavior of the individual should change. For instance, if a human root is identified as "decision to curb the scope of the formulary" (no name necessary), then the actions to correct that situation might be to provide the individual with training on how that decision can affect patient safety and to implement a series of checks and balances that would require a series of sign-offs for such a drastic decision to be made. These countermeasures will address the concerns of the human root without making a specific human root recommendation and giving the potential perception of a "witch hunt."

Responsibility of Executing the Recommendation

The responsibility of executing the recommendation will also be listed to identify an individual or group that shall be accountable for the successful implementation of the recommendation.

Estimated Completion Date

The estimated completion date provides an estimated time line for completion of each countermeasure, thus setting the anticipated time line of ROI.

Detailed Plan to Execute Recommendation

This section is generally an expansion of the root cause action matrix described above. All the economic justifications, plans to resource the project (if required), the funding allocations, etc. are located here.

ROOT CAUSE ANALYSIS

5 Ps Data Collection Strategies

The 5 Ps data collection strategies show the structured efforts to gain access to the necessary data to make the RCA successful through evidence-based verifications.

Logic Tree

The logic tree serves as a dynamic expert system (or troubleshooting flow diagram) for future analysts. This type of information will optimize the effectiveness of any corporate RCA effort by conveying valuable information to other facilities that have the potential for similar events.

Verification Logs

The verification logs are the spine of the logic tree and a vital part of the report. This section will house all of the supporting documentation for hypothesis validation. This is the actual evidence-based section of the report.

ANALYSIS TEAM STRUCTURE

Recognition of All Participants

Recognition of all participants is important if the intent is to have team members participate on RCA teams in the future. Every person who inputs any information into the analysis in this section should be noted, because people appreciate recognition for their successes.

Team Charter

The team charter shows the structure and focus that the team has displayed in their efforts.

Team Critical Success Factors

The team critical success factors show that the teams had guiding principles and defined the parameters of success.

Recommendation Acceptance Criteria (If Applicable)

The recommendation acceptance criteria should be listed to show that the recommendations were developed around a stated criteria provided by upper management. This can be helpful in explaining why certain countermeasures were chosen over others.

BUSINESS INFORMATION (OPTIONAL)

This section of the report is optional for various reasons. Typically, a healthcare facility does not want to report financials outside of their facility. Therefore, if an RCA report were being developed to comply with TJC requirements, this information would not generally be listed. When seeking to comply with any such regulatory requirements, it is common practice to provide only what is required. However, when the report is to be used internally and perhaps the CEO and/or the CFO will be on the distribution list, detailed business information may be appropriate. When such documents are developed, they should be covered under applicable peer review protections so that they are not discoverable.

Data Collection Costs

This is an aggregated cost of all the time and expenses of team members to collect failure data.

Hypothesis Verification Costs

This is an aggregated cost of all the time and expenses of team members to conduct tests to verify hypotheses.

Recommendation Development and Execution Costs

This is an aggregated cost of the time to develop recommendations and their projected costs to execute.

Projected Recommendation ROIs

This is a listing of projected ROIs for each of the recommendations proposed.

Projected RCA ROI

This is a projected ROI for conducting the RCA itself. The report will serve as a living document. If a corporation wishes to optimize the value of its intellectual capital using RCA, then issuing a formal professional report to relevant parties is necessary. Serious consideration should be given to RCA report distributions. Analysts should review their findings and recommendations and evaluate who else in their organization may have similar operations and therefore similar problems. These identified individuals or groups should be put on a distribution list for the report so they are aware that this particular event has been successfully analyzed and recommendations have been identified to eliminate the risk of recurrence. This optimizes the use of the information derived from the RCA.

In an information era, instant access to such documents is a must. Most corporations have their own internal intranets that provide an opportunity for the corporation to store these newly developed "dynamic expert systems" in an electronic format, allowing instant access. Corporations should explore the feasibility of adding such information to their intranets and allowing all facilities to access the information.

Whether the information is in a paper or electronic format, the ability to produce RCA documentation quickly could help some organizations from a legal standpoint. Whether it is a government regulatory agency, corporate lawyers, or insurance representatives, demonstrating that a disciplined RCA method was used to identify root causes can prevent some legal actions against the corporations as well as prevent fines from being imposed due to noncompliance with regulations. Most regulatory agencies require a form of RCA to be performed by the organizations but do not delineate the RCA method to be used. They just want to ensure that one can be demonstrated upon audit.

THE FINAL RCA PRESENTATION

This is the RCA team's final "day in court." It is what the entire body of RCA work is all about. Throughout the analysis, the team should be focused on this meeting. The analogy of the detective has been used throughout this text. In Chapter 7, the discussion focused on why detectives go to the lengths that they do in order to collect, analyze, and document data. The conclusion was that they knew they were going to court, and they knew the prosecution lawyers must present a solid case in order to obtain a conviction.

The analysis team's situation is not much different. Their "court" is a final executive review committee who will decide if their case is solid enough to approve the requested monies for implementing proposed recommendations.

Realizing the importance of this meeting, the team should prepare accordingly. Preparation involves the following steps:

Have the professionally prepared reports ready and accessible.
Strategize for the meeting by knowing the audience.
Have an agenda for the meeting.
Develop a clear and concise, professional presentation.
Coordinate the media to use in the presentation.
Conduct "dry runs" of the final presentation.
Prioritize recommendations based on impact and effort.
Determine "next-step" strategy.

Each of these steps is addressed below.

HAVE THE PROFESSIONALLY PREPARED REPORTS READY AND ACCESSIBLE

At this stage, the reports should be ready, in full color, and bound. Have a report for each member of the review team and a copy for each team member. Part of the report includes the logic tree development. The logic tree is the focal point of the entire RCA effort and should be graphically represented. The logic tree should be printed on large-scale paper, in full color, and laminated if possible (or printed on high-gloss paper). A large copy of the logic tree should be displayed on the wall in full view of the review committee. Keep in mind that this logic tree will allow for the executives to show other units, departments, and organizations how progressive their facility is in conducting RCAs. It will truly serve as a trophy for the organization.

STRATEGIZE FOR THE MEETING BY KNOWING THE AUDIENCE

This is an integral step in determining the success of the RCA effort. Many people believe they can develop a top-notch presentation that will suit all audiences. This has not been our experience. All audiences are different and therefore have different expectations and needs.

Consider the courtroom scenario again. Lawyers are courtroom strategists. They will build their case based on the makeup of the jury and the judge presiding. When the jury has been selected, they will determine their backgrounds (i.e., are they middle class, upper class, etc.). What is the ratio of men to women? What is the ethnic makeup of the jury? What is the judge's track record on cases similar to this one? What have previous cases that the judge ruled on been based on? Take this same scenario into the RCA presentation and you can see the importance of learning about the people you hope to influence.

In preparing for the final presentation, determine who the attendees will be. Then learn about their backgrounds. Are they clinical people, technical people, finance people, or business people? Making a clinical presentation to a financial group would risk the success of the meeting.

Next it must be determined what makes these people "tick." What are their incentives? Is it based on stock price, cost reduction goals, profitability, various ratios, and/ or patient safety records? This is important because the benefits of implementing the recommendations of the report must be expressed in units that appeal to the audience. For example, "If we are able to correctly implement the streamlined procedure for using Lovenox, then we will be able to cut current extended lengths of stay for applicable patients by 66%, which equates to an average savings of about $4100 per patient and a projected $7.0 million savings in the first year for the hospital as a whole!"

Have an Agenda for the Meeting

No matter what type of presentation you have, always have an agenda prepared. Management expects this formality, and it highlights the organizational skills of the team. Figure 10.2 is a sample agenda that we typically follow in our RCA presentations.

Always follow the agenda. Only break away when requested by the executive team. Notice that the last item on the agenda is "commitment to action." This is an important agenda item. Presenters sometimes leave meetings with a feeling of uncertainty, and they turn to their partners and ask, "How do you think it went?" A great deal of work has been done to get to this point, and the team members should not have to wonder how it went. It is not impolite or too forward to ask at the conclusion of such a meeting, "Where do we go from here?" Even a decision to do nothing is a decision, and people will know where things stand. Never leave the meeting wondering how it went.

Develop a Clear and Concise, Professional Presentation

Research shows the average attention span of individuals in executive positions is about 20 to 30 minutes. The presentation portion of the meeting should be designed to accommodate this time frame. The entire meeting should last no more than one hour. The remaining time will be left to review recommendations and develop action plans.

The presentation should be molded around the agenda developed earlier. Typical presentation software such as Microsoft's PowerPoint® provides excellent graphic capabilities and also easily allows the integration of digital graphics, animation, and sound clips. Remember this is the big "day in court," and all of the bells and whistles must be brought out to sell the recommendations. The use of various forms of media during a presentation provides an interesting forum for the audience and aids in their retention of the information.

#	Agenda Topic	Speaker
1	Review of PROACT Process	RJL
2	Summary of Undesirable Event	RJL
3	Description of Error Chain Found	KCL
4	Logic Tree Review	KCL
5	Root Cause Action Matrix Review	WTB
6	Recognition of Participants Involved	WCW
7	Question and Answer Session	ALL
8	Commitment to Action/Plan Development	RJL

FIGURE 10.2   Sample final presentation agenda.

There is a psychology behind how the human mind tends to react to various colors. This type of information should be considered during presentation development. The use of easel pads and LCD projectors provide a couple of different media to spice up the presentation. Props such as monitoring strips, patient records, failed equipment, or pictures from the incident can be passed around to the audience to enhance interest and retention. All of these things increase the chances of acceptance of recommendations.

COORDINATE THE MEDIA TO USE IN THE PRESENTATION

As discussed above, many forms of media should be used to make the presentation. Coordination of the use of these items should be worked out ahead of time to assure proper flow of the presentation. Lack of preparation could affect results of the meeting and result in a disconnected or disorganized presentation.

Tasks should be assigned prior to the final presentation. Such assignments may include a person to display overheads while the other presents, a person to manipulate the computer while the other presents, a person to hand out materials or props at the speaker's request, and a person who will provide verification data at the request of management. Such preparation and organization really make the team shine during a presentation and are apparent to the audience.

It is important to understand the layout logistics of the room that the team will be presenting in. Nothing is worse than showing up in a conference room and realizing that the laptop we brought is not compatible with the LCD projector provided. Then valuable time is spent fidgeting with equipment and trying to make it work in front of the audience. Some things to keep in mind are the following:

- Know how many people will be in the audience and where they will be sitting.
- Use name cards if you wish to strategically place certain people in certain positions around the table.
- Ensure that everyone can see the presentation from where they are sitting.
- Ensure that there is enough handout material (if applicable).
- Ensure that the A/V equipment is fully functional prior to the meeting.

Like everything else about RCA, the team must be proactive in their preparation for the final presentation. If the team blows it in the "courtroom," then the RCA cannot be successful, and patient safety efforts will be the loser.

CONDUCT "DRY RUNS" OF THE FINAL PRESENTATION

The final presentation should not be the testing ground for the presentation. No matter how prepared the team members may think they are, they should realize there is a possibility that there may be holes in their presentation and logic.

At least two "dry runs" should be conducted prior to the final presentation. These dry runs should be presented in front of the best constructive critics in the organization. These people will be happy to identify logic holes, therefore strengthening the logic of the tree. The time to find gaps in logic is prior to the final presentation, not during. Logic holes found during the final presentation will damage the credibility of the entire logic tree. This is a key step in preparation for the final presentation.

PRIORITIZE RECOMMENDATIONS BASED ON IMPACT AND EFFORT

Part of getting what the team wants from a presentation involves presenting the information in a "digestible" format. For instance, if an RCA resulted in 19 recommendations, now the task is to get these recommendations implemented. If 19 recommendations are put on the executive's desk, then there is a reduced likelihood that any will actually get done. Therefore the team must present them in a "digestible" manner. They must be presented in a format that appears to minimize the actual amount of things to be done. How can this be accomplished?

This is where a prioritization tool such as an impact-effort priority matrix can come in handy. This is a simple 3 × 3 table with the x-axis indicating "impact" and the y-axis indicating "effort to complete." Figure 10.3 is a sample of such a table.

Reflect back to the previous scenario of proposing 19 recommendations to an executive group. The team can separate the recommendations they have direct control to execute, and determine them to be high-impact, low-effort recommendations. The team can deem other recommendations as requiring other departments'

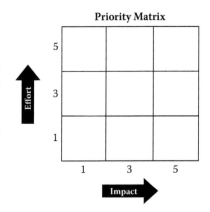

FIGURE 10.3 Impact-effort priority matrix.

approval; therefore, they may be a little more difficult to implement. Finally, the team may determine that some recommendations require corporate approval because of capital funds needed before any corrective action can be taken, making these recommendations more difficult to implement. This is a subjective evaluation that breaks down a perceived impossible pile into manageable and accomplishable tasks. A completed matrix may look like Figure 10.4.

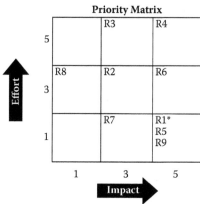

* Denotes a Recommendation

FIGURE 10.4 Completed impact-effort priority matrix.

DETERMINE NEXT-STEP STRATEGY

The ultimate result the team is looking for from this step (communicate findings and recommendations) is a correction action plan. This section deals with selling the recommendations and gaining approval to move on them. At the conclusion of the meeting, the team should have some recommendations that have been approved, individuals assigned to execute them, and time lines by which to complete them. The next phase will be the effectiveness of implementation and overall impact on bottom-line performance.

11 Tracking for Results

Consider what the team has accomplished thus far in the RCA process:

- Established management systems to support RCA
- Conducted an FMEA or opportunity analysis (OA)
- Developed a data preservation strategy
- Organized an ideal RCA team
- Utilized a disciplined method to determine accurate root causes
- Prepared a formal RCA report and presentation for executive management

Up to this point, this has been an immense amount of work and accomplishment. However, success is not defined as identifying root causes and developing recommendations. Patient safety has to improve as a result of implementing the recommendations!

Analysts must constantly show the organization why proactive technologies such as FMEA, OA, and RCA are fiscally responsible activities that can dramatically improve patient safety. Tracking for results becomes the team's measurement of their success using RCA methods (not merely regulatory compliance!). Therefore, since this is a reflection of the team's work, they should be diligent in measuring their progress because it will be viewed as a report card of sorts.

Once successes are established, they should be exploited by publicizing them for maximum personal and organizational benefit. The more people who are aware of the success of the proactive efforts, the more they will depend on them to eliminate future problems. This makes analysts a valuable resource to the organization. The more successful RCA becomes, the more rewards RCA teams will get, as various departments or areas will request the RCA service from those deemed as the experts. While this is a good indication of others jumping on board, there can be drawbacks.

For instance, we have shown that the analysts should work on the "significant few" events that make up about 80% of the organization's losses. Analysts may be asked to solve peoples' "pet problems" (not included in "significant few" list) that are not necessarily important to the organization as a whole. When this type of request is declined, the analyst may be viewed as not being a team player because he insists on sticking to the "significant few" list from the FMEA or the OA. These legitimate concerns about overcoming this potential issue should be discussed with the champions and drivers.

Let's pick up from the point where management has approved various recommendations of ours in our final meeting. Now what happens? We must consider each of the following steps:

- Sliding the proactive work scale
- Developing metrics to track

- Exploiting successes
- Creating a critical mass
- Recognizing the life cycle effects of RCA on the organization

SLIDING THE PROACTIVE WORK SCALE

The most common objection to performing RCA is that people simply do not have the time to do it properly. When this objection is looked at introspectively, it is found that people do not have the time because they are too busy "firefighting" or tending to the reactive needs of the organization. This truly is an oxymoron. RCA is designed to eliminate the need to put out fires. Executives should understand this and include RCA as part of the overall patient safety strategy.

Proaction and reaction should be inversely proportional. The more proactive tasks performed, the less reactive work there should be. Therefore, all the personnel we currently have conducting strictly reactive work will now have more time to face the challenges of proactive work. No healthcare facility has ever reported to us that they have all the resources they would like to conduct proactive tasks such as inspections, FMEAs, opportunity analyses (OAs) and RCAs. Organizations do not have these resources now because they are all fire fighting. As the level of proaction increases, the level of reaction will decrease. This is a point where control is gained of the operation and the operation does not control us (see Figure 11.1)!

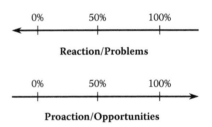

FIGURE 11.1   Inverse relationship between reaction and proaction.

DEVELOPING TRACKING METRICS

Recognizing that inverse relationship between proaction and reaction, the focus must now be on how to measure the effects of implemented recommendations. This is generally not a complex task, because typically there was an existing measurement system that identified a deficiency in the first place. By the time the RCA has been completed and the causes all identified, the metric to measure usually becomes obvious.

Below is a review of a few circumstances to determine appropriate metrics.

HEALTHCARE EQUIPMENT: IV PUMP FAILURES

A hospital experiences a mean time between failure (MTBF) of about 6 months on a certain manufacturer's new IV pump. Some of the root causes identified in the analysis uncover that many of the modes for failure were caused simply because the nurses did not understand how to use the new equipment. The new equipment was not as user-friendly as the old equipment. This was not a consideration in the purchase decision, as the new equipment was less expensive in the short term.

Understanding this, recommendations were implemented to provide users of this new equipment with basic manufacturer's training in the operation of the new IV pump. When future considerations were made for such purchases, lifecycle costs associated with the product would be the driving consideration instead of simply initial cost. As a result of this, the reported malfunctions of the equipment should decrease significantly (see Figure 11.2).

ED: ASPIRIN DELAYS

A hospital ED is audited by its insurance company. The insurance company offers higher reimbursement incentives for patient safety initiatives that meet certain patient safety goals. One of these is how often potential heart attack patients receive aspirin as soon as they enter the ED. The hospital gets "poor" ratings because records indicate they offer such aspirin in less than 80% of such cases. Under the incentive program, they get "0" points for this. The RCA reveals that this is simply an awareness and education issue for the triage nurses and the ED physicians. Awareness training is provided. The primary tracking metric will now be to ensure that the highest incentive is attained for providing the aspirin over 95% of the time (see Figure 11.3).

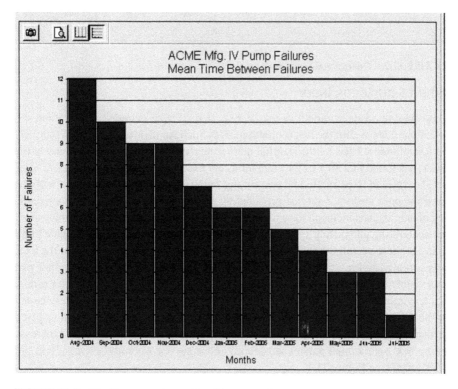

FIGURE 11.2   Healthcare equipment tracking example.

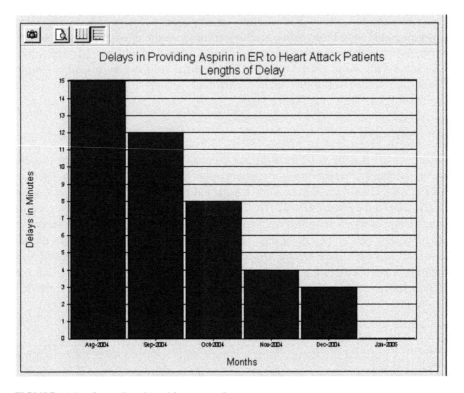

FIGURE 11.3 Operational tracking example.

PATIENT SATISFACTION INDEX

A hospital in a very competitive market experiences a 5% decrease in its patient satisfaction index within the last 6-month period. Such surveys are conducted routinely on a semiannual basis by the hospital marketing department. Upon conclusion of the RCA, it is found that 80% of the complaints are associated with billing issues. Causes are determined to be inefficient accounting practices, poor communication with the patients, poor customer service provided by the billing department, and poor tracking of actual costs incurred for a particular patient. As a result, a separate analysis on the accounting and billing system reveals several "holes." Attempts were made to fill many of these accounting holes via recent fixes, but poor implementation of the fixes was actually making things worse than before, and this surfaced in a manner that was visible to the patient. Efforts were quickly taken to plan and schedule for proper implementation of the recommendations in a timely manner. Apologies were issued to the patients affected in the past, and their concerns were addressed and resolved immediately. A key metric that can be used to measure success will be an increase in the next semiannual patient satisfaction index survey where billing concerns are questioned (see Figure 11.4).

The pattern of metric development described above shows that the metric initially indicated that something was wrong. This metric can also be (and usually is) the same lead metric that indicates something is improving. Sometimes determin-

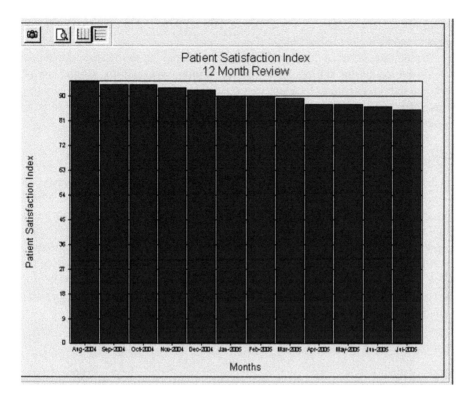

FIGURE 11.4   Customer service tracking example.

ing the appripriate metrics seems too simple, and therefore it cannot be if it is that simple. As a result, the "paralysis by analysis" paradigms set in, and the analyst develops complex measurement techniques that can be overkill. This is not to say they are never warranted, but efforts should be taken not to overcomplicate issues that do not require such complexity.

EXPLOITING SUCCESSES

If no one knows the successes generated from RCA, then the initiative will have a hard road ahead of it, and the organization will not be optimizing the effects of the analyses. Like any new initiative in an organization, skepticism abounds about its survival chances. As discussed earlier about the "program-of-the-month" mentality, skepticism is likely to set in soon after the introduction of any initiative. To combat this, RCA successes should be exploited to improve the chances that the initiative will remain viable and accepted by the user population. Without this participation and acceptance, the effort is typically doomed.

How do we effectively exploit successes? One of the main ways is through high-exposure media, including such final RCA report distributions as internal hospital newsletters, corporate newsletters, company intranet, presentations at trade conferences, articles for trade publications and journals, and finally, exposure in texts such as this for successful outcomes demonstrated through the use of case studies. These types of media are described below.

Exploitation serves a dual purpose: it gives recognition to the corporation as a progressive entity that utilizes its workforce's brainpower, and it gives the analyst and core team recognition for a job well done. This is the motivator for continuing to perform this work. Without recognition, analysts will tend to move on to other things, because there is no personal satisfaction in this type of work.

REPORT DISTRIBUTION

As discussed in Chapter 10, to optimize the impact of RCA, the results must be communicated to the people who can best use the information. In doing this, it is also important to communicate to these people that the analysts are doing some pretty good work in the name of RCA and that their people are being recognized for it. It is important to include the proper executives in the report distribution. This is because RCA is often seen in a very narrow view. Normally, it is viewed as the task of the risk manager or the quality manager. However, executives need to learn that RCA is everyone's job, and it is applicable anywhere, from the administrative offices to the patient on the floor and to the engineers in facilities.

INTERNAL NEWSLETTERS

Most organizations have some type of an internal newsletter. These newsletters serve the same purpose as a local newspaper—to communicate useful information to its readers. Most publishers of internal newsletters welcome such success stories for use in their newsletter. That is what newsletters are about. Therefore, take advantage of the opportunity. Patients will read this as well and understand the progressive manner in which the facility is working to ensure patient safety.

CORPORATE NEWSLETTERS

Most organizations have some type of corporate newsletter. It may not be published as frequently as the internal newsletter, but it is published periodically. These types of newsletters typically focus on the "big" picture as compared to the internal newsletter and may include more articles geared toward financial information, competition in their market, etc. However, they too are looking for success stories that can demonstrate how their corporation increased patient safety and saved money to boot. Such newsletters also provide executives with bragging rights when talking to their peers in the industry.

PRESENTATIONS AT TRADE CONFERENCES

This is a great form of recognition for both the individual (and team) and the organization. For some analysts, this is their first appearance in a public forum. While some people may be hesitant about the public speaking aspect of the event, the ability to get through it and receive the applause of an appreciative crowd can dispel any previous misgivings. Trade conferences thrive on the input of the companies involved in the conference. They are made up of such successes, and the conference is a forum to communicate valuable information to others who can learn from it.

ARTICLES IN TRADE PUBLICATIONS

Trade publications provide exposure to possibly thousands of individuals in the publication's circulation. Reprints of these articles tend to be viewed as trophies to the analysts, who are not used to such recognition. When we have such star client analysts who have written an article about their successes, we frame the reprint and send it to the analyst for display in the office. It is a career accomplishment the analyst should be proud of, and it also looks good on a resume!

CASE HISTORIES IN CLINICAL JOURNALS

Healthcare journals are typically peer-review journals. This means that these publications submit articles and case studies to a peer review committee that seeks to validate the content of the work and establish the credibility and interest the piece will have to the readers of the journal. Because RCA is a system based on fact, it is an honor to publish in such journals where a group of peers has the opportunity to see the depth of the analyst's work. Analysts gain credibility and status if they are able to publish in such vetted journals.

CREATING A CRITICAL MASS

When discussing the term "critical mass," reference is made not only to RCA efforts but the introduction of any new technology. It has been our firm's experience in training and implementation of RCA efforts over the past 30 years that, if a critical mass of 30% of the potential analysts on board can be created, then the others will follow.

The "program-of-the-month" mentality has been beaten to death, but it is reality. Some people are leaders and others are followers. The leaders are generally the risk takers and the ones who welcome new technologies to try out. The followers are typically more conservative people who take a "wait and see" attitude. These people believe that if this is another "program-of-the-month," they will wait it out to see if it has any staying power. They have been hyped up before about such new efforts, possibly even participated, and then never heard any feedback about their work. They are alienated with regard to "new" thinking and the seriousness of executive management to support it.

If 30% of the population trained in RCA were actually encouraged to use their new skills in the hospital to increase patient safety and produce bottom-line results, then RCA would become more institutionalized in the organization. If only 30% of the analysts started to show financial results, the dollars saved would be phenomenal. Executives would then continue to support the effort with actions, not words. Once the analysts start to get recognition within the organization, others will crave similar recognition and want to participate.

It is unrealistic to expect that everyone trained will respond in the manner the organization would like. It *is* realistic to expect a certain percentage of the population to take the new skills to heart and produce results that will encourage others to come on board.

RECOGNIZING THE LIFE-CYCLE EFFECTS OF RCA ON THE ORGANIZATION

RCA can play a major role in today's overall corporate strategies for growth. As referenced throughout this text, the goal should be elimination of the risk of recurrence of any undesirable outcomes. Many organizations set their sights on being the best "predictors" of such events and thus target the reduction in response time as the successful measure. While this is still a must in the interim, it should be a means to another end: elimination of the recurrence. If undesirable outcomes did not exist, there would not be a need to become better predictors.

Millions of dollars have been spent by healthcare organizations on Six Sigma approaches. As discussed in Chapter 6 (comparing the goals of Six Sigma to RCA), these philosophies should be melded together to create a comprehensive corporate patient safety strategy.

Why is it that millions of dollars are being funneled into Six Sigma approaches (whose goal is to minimize process variation or mitigate consequences) and very little attention is given to RCA (whose goal is to eliminate the risk of recurrence of an adverse outcome)?

While many organizations have grasped the Six Sigma approach, analysts continue to have difficulty convincing corporate executives to give equal credence to an RCA effort. When implemented appropriately, RCA is eliminating the chances of a current failure from recurring, and these failures are even being compensated for (accommodated) in the budget. When chronic issues are solved and eliminated, there is no need to budget for their occurrence any longer. The savings are off the bottom line in the same fiscal year.

Organizations have been inundated with what is referred to as "zero imperatives" (Figure 11.5). Zero imperatives are the efforts associated with zero tolerances. These consist of zero inventory (just-in-time), zero injuries, zero quality defects (TPM), etc. RCA is geared toward zero failures, or the elimination of undesirable events.

Realistically, we realize that these zero imperatives are not likely to be 100% obtainable, but they do provide something to strive for. Six Sigma, from a statistical standpoint (3.4 defects per

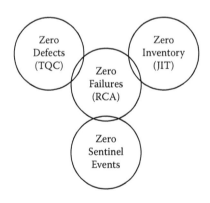

FIGURE 11.5   The zero imperatives.

99% Good (3.8 Sigma)	99.99966% Good (6 Sigma)
20,000 lost articles of mail per hour	Seven articles lost per hour
Unsafe drinking water for almost 15 minutes each day	One unsafe minute every 7 months
5,000 incorrect surgical operations per week	**1.7 incorrect operations per week**
Two short or long landings at most major airports each day	One short or long landing every 5 years
200,000 wrong drug prescriptions each year	68 wrong prescriptions per year
No electricity for almost 7 hours each month	One hour without electricity every 34 years

FIGURE 11.6   Precision comparison.

million opportunities), exemplifies this point very well in its strive for process precision (see Figure 11.6).*

CONCLUSION

Human efforts will never be error-free, but they can strive to be. Precision is a state of mind and requires the mentality to constantly strive for the next plateau.

PROACT® RCA as described in this text is not a panacea. It is merely a method to assist in logical thinking to resolve undesirable events. While many of the analogies have been from the healthcare world, this PROACT RCA approach is applicable under any circumstances. Whether it is chronic or sporadic, mechanical or administrative, in an oil refinery or a hospital, all require the same logical human thought process to resolve their respective issues.

In the following chapters we will discuss how to make this thought process more manageable. Efforts will be taken to alleviate the administrative burden of managing an RCA by providing a simple and effective software solution. While conducting RCA in a disciplined manner can be difficult, most of the time is spent in sticking to the discipline and documenting the process. One of the ways an incentive can be provided is to take this extra step of "discipline" to make the task easier and more desirable to do. This is where the PROACT Suite plays its role.

This essentially ends the textual and philosophical portions of the book. From this point on, we will discuss how to make our analysis efforts more efficient (and less stressful) via the use of the PROACT Suite software tools designed for the sole purpose of conducting FMEAs, OAs, and RCAs.

---

* Grinnan, M.D., Richardson, Quality-In-Sights®: Hospital Incentive Program (Anthem Q-HIP Program), VASHRM Meeting presentation, Charlottesville, VA, 2003.

V

*A Software Technology
to Support Proaction*

12 Automating Proactive Analyses

The Utilization of the PROACT® Suite Software Solution (Version 3.0+)

PROACT® is the acronym that has been referenced in this text to describe a root cause analysis (RCA) methodology. In this chapter, PROACT (Version 3.0+) will be used to describe the name of a software package that helps facilitate an RCA as well as a FMEA and OA. We will relate how and where opportunities exist to automate tasks that are otherwise done manually in the performance of an RCA.

We have discussed at length the pros and cons of conducting an FMEA and OA, both manually and automated, using various sources of data. Data generally come from current computerized management systems (CMSs) institutionalized within organizations, such as incident management systems (IMSs), as well as directly from the raw source—people.

Reliance on this data is a good starting; however, it must be realized that "sleepers" exist. Sleepers are the tasks that happen so often they are typically not recorded in the IMS. This is because these sleepers characteristically do not take a great deal of time to address or correct. It is seen as a waste of time to enter them into the system, because the entry could take longer than the fix did. The problem comes when such sleepers happen 500 times a year and no recording mechanism picks them up. This is generally what the operations budget accepts as a "cost of doing business."

The point is to make sure that the IMSs are not viewed as the cure-all for RCA problems. Until organizations are 100% confident that the CMS is truly reflective of the activity on the floors, they should consider the use of interviews with the people who do the work as the greatest source of data for an FMEA and OA.

CUSTOMIZING PROACT FOR A FACILITY

One important feature of RCA software is that it is customizable for each specific facility. The software accommodates site information (facility locations, divisions, and departments) and medical equipment listings (type and class). This makes it possible for analysts to choose from a list of items specific to their particular facility.

Part of the initial setup in PROACT will also allow the user to import a human resources listing of active personnel whom the facility would like to include as potential team members for an RCA. These individuals can be chosen on an as-needed basis for each individual analysis.

SETTING UP A NEW ANALYSIS

Once the facilities and medical equipment information have been stored, the analyst can set up a new analysis. PROACT will carry users through a setup wizard, which will ask them to input certain base information about the analysis. This base information is stored in a database as well for use with search features to be described later on. The *New Analysis Wizard* involves a series of six steps.

STEP 1

Step 1 is the inputting of the analysis name, description, and type (Figure 12.1). The type is very important. The program permits the user to pick the analysis type from a list, citing such choices as safety, risk, quality, mechanical, environmental, operational, and administrative. This listing is also configurable per the user's requirements. This will allow the eventual sorting of the analysis database based on these categories. Therefore, when risk managers (RMs) want to view all of the completed analyses on risk issues, they can simply sort the database on this field.

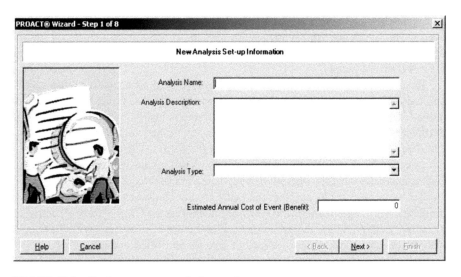

FIGURE 12.1   Setting up a new analysis, step 1.

STEP 2

This step (Figure 12.2) involves identifying the specifics about the event being analyzed. This will assist the analyst later when trying to "data mine" a database for information about specific events that have been analyzed.

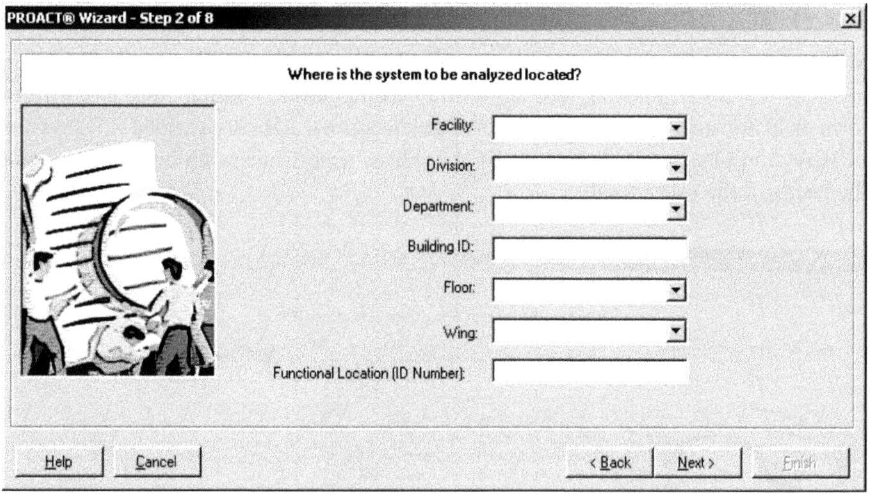

FIGURE 12.2   Setting up a new analysis, step 2.

STEP 3

Step 3 (Figure 12.3) will involve the setting up of team members for the specific analysis at hand. Again, a prepopulated database will exist with the facility's personnel. The analyst will then be able to choose who would be most suitable for this analysis team.

Step 3 will also allow the principal analyst (PA) the latitude to grant permission levels to each team member. Permissions include the following: (1) read only, (2) read/write, and (3) delete. This comes in handy when the team may include temporary or contract employees and the PA does not wish to grant full access.

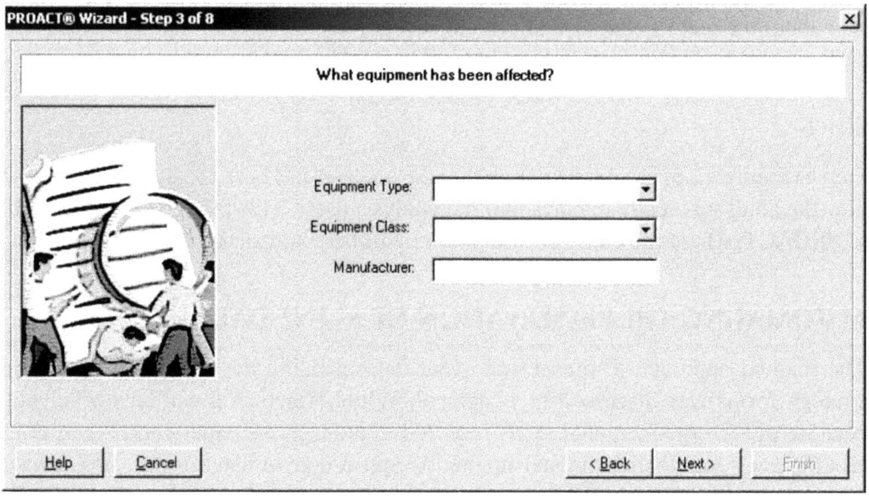

FIGURE 12.3   Setting up a new analysis, step 3.

STEP **4**

Step 4 (Figure 12.4) will involve setting up the team information. This information includes when the analysis will start, when completion is anticipated, and a free-form field for any general comments. Predetermined defaults include 45 days for FMEAs and OAs, and 30 days for RCAs. These time frames can be altered to suit the needs of the user facility.

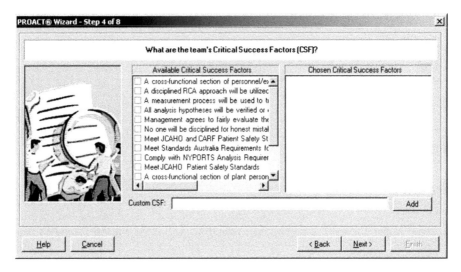

FIGURE 12.4   Setting up a new analysis, step 4.

STEP **5**

Step 5 (Figure 12.5) involves setting up specific team information. As discussed in Chapter 8 a direction or focus for the team should be planned. This is where information about the team's charter and critical success factors (CSFs), that was discussed earlier, would reside.

STEP **6**

Step 6 (Figure 12.6) is a confirmation that the analysis has been created successfully; now the analyst is ready to enter into the analysis itself. This takes us into the "PR" of PROACT, where the data collection tasks will be automated.

AUTOMATING THE PRESERVATION OF EVENT DATA

The manual approach to preserving event data utilizing the 5 Ps data collection strategy forms was discussed in Chapter 8. While effective, it can lack efficiency because of the organizational skills required to manage the paperwork. Also, from an efficiency standpoint, manual methods require double handling of data, which is non-value added work. Whenever information is written down, it will eventually

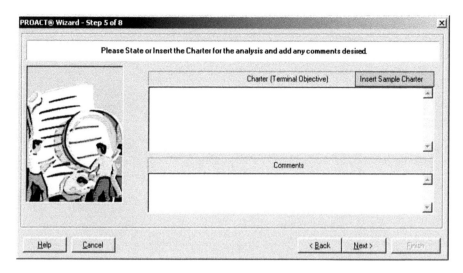

FIGURE 12.5   Setting up a new analysis, step 5.

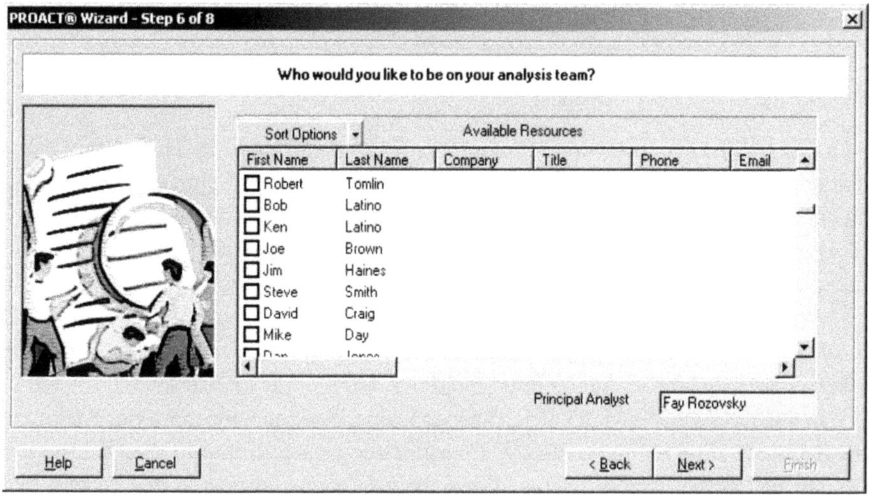

FIGURE 12.6   Setting up a new analysis, step 6.

have to be entered into a computer for final presentation. Automation provides an opportunity to eliminate one step.

Now the tab designated PRESERVE will be opened, and the analyst will see a screen with the same fields as on the manual form, except they are now allowing a database to be developed in the background (Figure 12.7). Notice that the five icons to the left represent the PROACT acronym. Selecting an icon opens up the screen associated with that step in the PROACT RCA process.

The first team meeting would involve a brainstorming session of the core team. The team would assemble and, based on the facts at hand, start to develop a list of

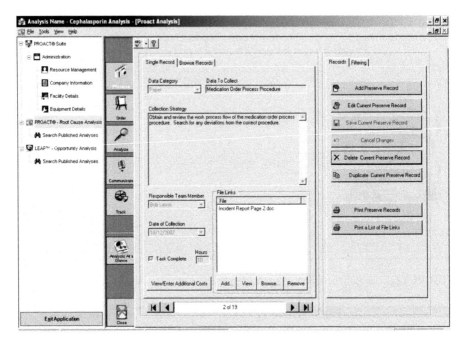

FIGURE 12.7   Preserve—opening screen.

data that will be necessary to collect in order to start the analysis. This type of automation is most effective if a laptop computer is available in the meeting room with an operator/recorder entering data as it is offered. The ideal situation is the use of an LCD projector with the laptop computer, so the entries can be seen on the screen and everyone can be assured their information is transcribed accurately.

As the type of data to collect is entered, a team member should be assigned to obtain information using a certain collection strategy. A time frame should be assigned to focus the team and forge the progression of the analysis.

In the PRESERVE MODULE, the user is provided the opportunity to document his data collection strategy. PROACT permits the user to link imported files to the desired PRESERVE record. This allows the user to document the record with pictures of failed parts, results of lab testing, applicable procedures, and so forth. This detail in documentation helps build our "solid case" with documented hard proof to back up our analysis (Figure 12.8).

Notice in the lower left corner the button labeled "View Additional Costs" (Figure 12.9). In order to be able to calculate projected returns on investment (ROIs), certain cost information must be collected. One of those costs is team members' time and expenses while conducting tasks associated with an RCA.

In PROACT, there are three locations that will allow the PA to assign tasks associated with the RCA: (1) data collection, (2) hypothesis verification, and (3) recommendation writing and implementation. Anywhere in the program where tasks are assigned, the analyst will be able to acknowledge that the task was completed and enter the total amount of time it took to complete the task.

Available File Links   _ □ ×

File Name	New Directory	Assign
☐ Broch Fire.bmp	C:\Program Files\RCI\PROAC'	
☐ Bronch Fire 3.bmp	C:\Program Files\RCI\PROAC'	
☐ bronchoscope_fire_mishap.pdf	C:\Program Files\RCI\PROAC'	Import...
☐ doctor.jpg	C:\Program Files\RCI\PROAC'	
☐ Incident Report Page 2.doc	C:\Program Files\RCI\PROAC'	Remove
☐ Lexington RN Report.doc	C:\Program Files\RCI\PROAC'	
☐ Neonate Tree.pdf	C:\Program Files\RCI\PROAC'	
☐ Progress Notes.doc	C:\Program Files\RCI\PROAC'	

Close

FIGURE 12.8   Preserve—file linking screen.

These hours will be multiplied by a pay rate that was established in the personal team member record. This pay rate information can be accurate to the penny or just categorized by general position. This sensitive information will be confidential and cannot be accessed unless permission is granted by the IT group in the form of a global or site administrator (Figure 12.10).

The software provides a filtering and sorting option that will allow the PA to print the lists at the end of the meeting in an array of formats. Sorts can be performed on various fields. Filters can be established to print only certain information and exclude all other. For instance, the analyst may want to print only one team member's assignments versus those of all team members. A browse feature is also available to see the entire list at a glance.

FIGURE 12.9   Preserve—view additional costs button.

PROACT provides the unique ability to send automatic e-mail notifications to people who have been assigned tasks (Figure 12.11). These reminders will be sent at predetermined intervals, which are set by the analysts. When working in very reactive environments, sometimes those assigned tasks are forgotten about because the worker is busy with other tasks. This is a nonpersonal and nonthreatening way to politely remind them of when tasks are due. This feature, like the "view additional costs" feature, is located in all areas of the software where people are assigned tasks. Using this automation throughout the process helps keep the team interested and organized.

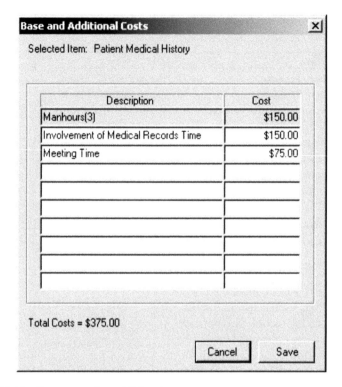

FIGURE 12.10   Preserve—view additional costs pop-up.

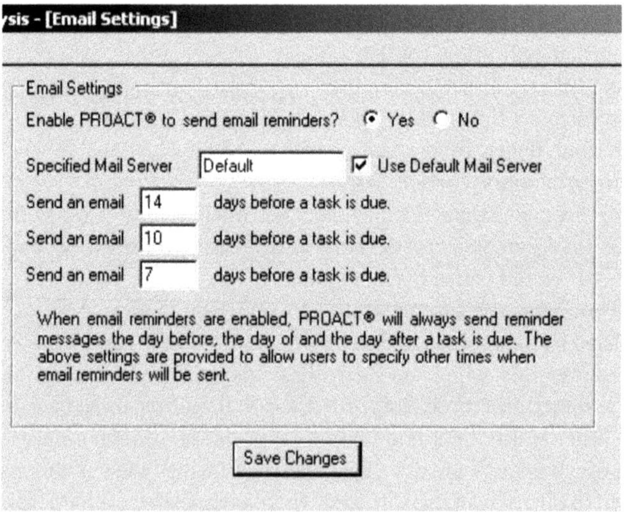

FIGURE 12.11   Preserve—review e-mail reminder.

AUTOMATING THE ANALYSIS TEAM STRUCTURE

The importance of the team structure has been discussed at length in this text. The diversity of backgrounds of the team members was a key consideration to obtaining a successful result. It was also stressed that the leader of an RCA team should typically not be the expert in the event being analyzed because of the inherent bias that may exist.

The focus of the team structure was discussed by formalizing the team entity through the development of a team charter and the identification of CSFs. These tasks show management that there was considerable thought about why the team was formed and what their objectives are in obtaining success.

Now let us contrast the manual approach versus the automated approach. In a manual format, the team would most likely be utilizing a paper filing system to record team member information. They would also likely use a word processing program to develop the team charter and the CSFs. With an automated format, PROACT can be used to catalog all this information in one location along with the 5 Ps information collected previously.

Remember, previously, all of these data were collected during the initial NEW ANALYSIS WIZARD (six steps) when the new analysis was created. The ORDER THE ANALYSIS TEAM tab is where that information is stored and available for modifications. Figure 12.12 shows how a change in granting permissions to a team member can easily be accomplished.

PROACT will maintain a team pool by which a database of qualified RCA team members is stored. A qualified team member may be a past RCA participant, an individual who has received RCA training in the past, or an individual who possesses a particular expertise pertinent to this analysis. In any case, maintaining a record of such people is an efficient way of organizing RCA teams. Once a reservoir of talent has been identified, then specific individuals can be assigned to lead and participate on the core team. These choices will vary based on the nature of the event being analyzed. PROACT will allow reports to be developed on the team members based on their names and/or telephone numbers (see Figure 12.13).

PROACT now has all the team information cataloged and organized within a database. Up until this point, there has been no need to utilize individual database or spreadsheet programs or word processing programs. It is all located within one RCA file.

AUTOMATING THE LOGIC TREE DEVELOPMENT

Assume the team is now at a point in the analysis where the initial data has been assigned and collected. Now the real issue of analyzing the data to determine what happened must be faced.

Using the manual method to develop the logic tree has its pros and cons. One of the disadvantages is that it is double handling of data. In the manual method, a logic tree might be built in a conference room where a mural has been put together, made of easel pad paper or craft paper. Subsequently, the analysts will facilitate the team using Post-It notes to record hypotheses on easel pads. This means that at some point

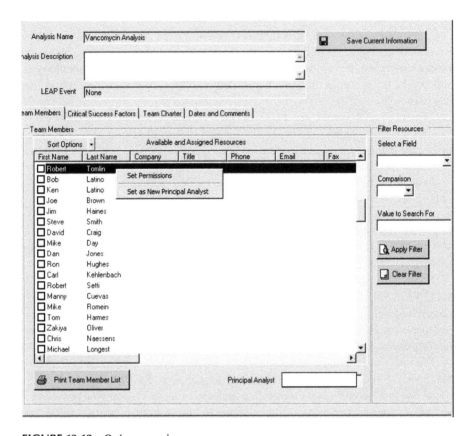

FIGURE 12.12   Order—opening screen.

in time, this information will have to be transcribed into another format for inclusion into the report and/or display in a presentation. This double handling leads to an inefficiency of time. When a team meeting ends, the team members usually do not have the updated logic tree until days later. This results in unnecessary delays before all team members have consistent information.

However, one of the psychological advantages of using the manual method in conjunction with the automated method is it may be perceived as accomplishing work. Most RCA teams have seen the paradigms at play where it is felt that if someone is working on a computer all the time, "work" is not being accomplished. The same can be said for RCA. If management walks by a conference room where an RCA team is meeting and only sees one laptop on the table and five team members sitting around talking, then it can be perceived as a non-value added use of time. However, in the same scenario, if a manager walks by and sees this huge craft paper on the wall with all these Post-It notes, then that can be deemed tangible work (even if someone has duplicated the logic tree within PROACT on the laptop within the same meeting).

From an efficiency standpoint, using a laptop and an LCD projector in a team meeting is the ideal forum to conduct logic tree building sessions. Whether to use

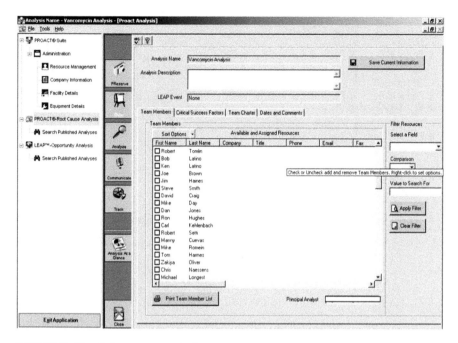

FIGURE 12.13 Order—team pool tab.

a laptop or manual devices will have to be the decision of the analyst or the team, based on the resources available at their site.

PROACT can make it a much more efficient process, thus automating the methodology. The PROACT software was developed using the same logic rules as discussed in the PROACT RCA methodology described in this text.

The opening screen in ANALYZE is basically a blank worksheet with the necessary tools or tabs to build the logic tree (Figure 12.14). The intent of this text is not to teach analysts how to use the software, but rather to make them aware of its capabilities. For this reason we will not describe all the features available— just the major ones that can make life easier for the analyst and the team. Each of the tabs and buttons on this screen will be briefly described here (Figure 12.15).

BUTTON FUNCTIONS

Top Box Wizard

When starting a new analysis, the team will have to identify the top box, which comprises the event and modes. This three-step wizard will ask for these inputs and automatically put them in position to start the logic tree. If an FMEA or OA had been conducted prior to this RCA, this information could be automatically bridged over from the other programs, making it easier for the analysis team.

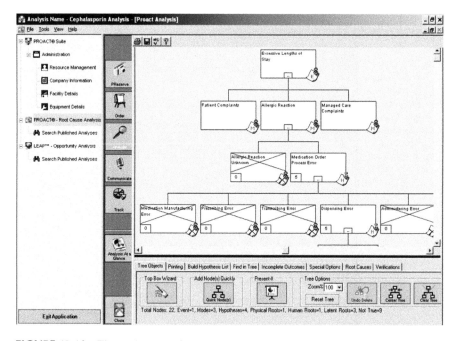

FIGURE 12.14   The analyze opening screen.

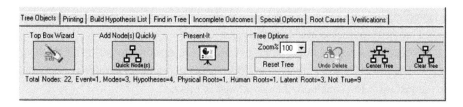

FIGURE 12.15   The analyze buttons and tabs.

Add Nodes Quickly

This is the primary button for adding hypotheses onto the logic tree. Users highlight the block under which they want to add a hypothesis (often referred to as the *parent*), then click this button, and it is placed there (this subordinate position is often referred to as the *child*). Then all an analyst has to do is type the text in the block added.

Present-It

This is one of the key features that saves the analyst an immense amount of time. Instead of having to develop a presentation using a graphics program, the presentation can be generated directly from PROACT with the tree that is already done. The software also has a great deal of versatility for presentations, as the blocks can be expanded and collapsed. When management questions how the team knows whether a hypotheses is true or not, the analyst can simply double-click on the block in question, and the verification details pop up. Verification details include the test used to prove the hypothesis, the outcome of the test, any file link supporting the hypotheses (i.e., pictures, reports, procedures, etc.), who did the test, when it was done, and how much time was spent proving the hypothesis.

Zoom

This allows the analyst to adjust the size of the logic tree for viewing by the audience.

Undo Delete

This feature allows the user to undo the last delete. We all are used to using this feature in other programs where we made an incorrect entry and wished to return to our "preclick" position. PROACT permits this to occur for the last delete.

Center Tree

This feature allows the analyst, anywhere in the tree, to automatically bring the tree to the top center of the screen.

Clear Tree

This button has an important and dangerous function in that it can delete the entire tree. In order to do this, a second screen will require another validation to execute the request to ensure that is what the analyst wants to do. This feature was put in to easily allow people who felt they got off on the wrong foot to scrap the old thinking and start over.

TAB FUNCTIONS

Printing

Printing items such as logic trees can be difficult in some programs, as the transition from page to page may not be clean and the lines may not connect. In this program, the user is offered the option of either (1) scale to one page or (2) best fit. Formatting is done automatically based on paper size and orientation chosen. The software displays the preview to the analyst to ensure that it is as desired. The user also has the option of using page connectors to orient one page with another when the logic tree transverses across multiple pages.

Build Hypothesis List

Those who have facilitated many RCA teams in the past know that the facilitator's focus must be on the process itself. When team members are asked *how something* could have happened, they will be throwing out hypotheses very quickly. It may break the train of thought to have to take one suggestion and then go through how to validate it, and then go on to the next one. The build hypothesis list feature allows the analyst to store all of these ideas in one location and then move them into the tree in the appropriate locations.

Find in Tree

An analyst who is looking on the tree to see if a particular word has been used already can type the word in and hit "search," and all the blocks containing that word will be acknowledged. This is also helpful when discussing the logic tree with a person in a remote location whom you want to quickly find a particular block on the logic tree.

Incomplete Outcomes

This is a helpful feature to the analyst in that it is basically an exception list. This report will identify whether all the blocks on the tree have (1) a verification test assigned or not and (2) whether the tests have been completed or not. This provides the analyst with a snapshot of the integrity of the logic tree at any given point.

Special Options

This tab provides numerous special functions where the logic tree is concerned. The user can opt to display a legend on the screen that identifies the various icons on the tree such as event (E), mode (M), hypothesis (H), not true (X), physical root (PR), human root (HR), and latent root (LR). Another helpful feature in this tab is the ability to show the "path to failure." A red line travels from the event to all identified causes, demonstrating the cause-and-effect relationships leading up to the undesirable outcome.

Sometimes users want to export the logic tree to other graphic programs like MS PowerPoint. PROACT now allows users to export the logic tree to aid in external presentations.

Lastly, in this location the user can add titles and subtitles to the printout of the logic tree.

Root Cause Listing

This is a report that will allow the user to print out a summary of the identified root causes only.

Verification Log

This is a report that will allow the user to print out a snapshot of all the verifications log items related to all the hypotheses listed. While the "incomplete outcomes" report shows only exceptions, this listing shows the current status of all verifications, whether complete or incomplete.

PROACT has a built-in knowledge management feature that allows the analyst to capitalize on the successful logic of past analysts. If a team develops a certain hypothesis and wants to see how other teams may have answered the same "how can?" question, it can use the "previous suggestions" feature. The team would simply highlight the word(s) to search in the published database, and a listing of previous suggestions would pop up. At this point, the team can choose the ones that are appropriate for its analysis, and the software will automatically place them on the logic tree as a child to the parent in which the search was conducted.

Using this feature, PROACT incorporates the TJC root cause action matrix categories into the database. For instance, if a "medication error" (a reviewable sentinel event) block has been chosen and the "previous suggestions" feature used, all of the TJC categories to check would pop up, and the team can select which are appropriate (Figure 12.16).

Much effort went into the design of PROACT to make it as user friendly as possible. When building the logic tree, there are numerous ways to complete the same task, depending on the preference of the user. For instance, the analyst can add a new hypothesis by (1) using the quick node feature, (2) clicking on the toolbar beneath the current hypothesis, or (3) using the add node feature when right-clicking on the mouse. The right-click menu has all the same options as the main function bar, for those who prefer to use their mouse for most of the tasks (Figure 12.17, Figure 12.18).

When the logic tree is completed, it is important to "dress it up" properly for final presentation to those who will have to approve the recommendations. The physical, human, and latent root causes will have to be labeled. This is easily accomplished by highlighting the block to receive the label then clicking on the toolbar beneath it and selecting the appropriate root cause designation. As mentioned earlier, this task can also be accomplished using the right mouse click menu (Figure 12.19).

Sometimes it is helpful to be able to show the audience the path to failure in a unique manner. PROACT allows this to be accomplished by highlighting the path using a thick red line leading to all of the hypotheses proven to be true and/or that

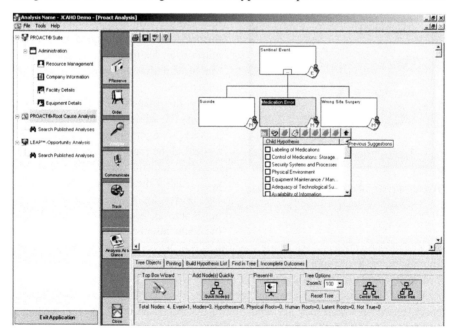

FIGURE 12.16   Previous suggestions feature using TJC matrix.

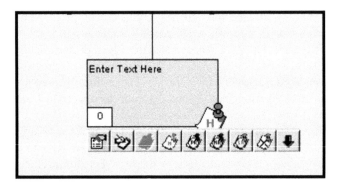

FIGURE 12.17   Hypothesis/node toolbar options.

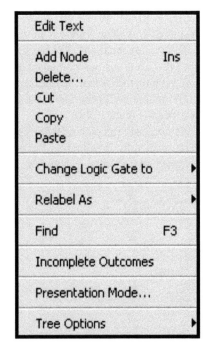

FIGURE 12.18   Right mouse-click menu.

result in designated root causes where recommendations are developed. This can be quite impressive for the final presentation of RCA results (Figure 12.20).

AUTOMATING RCA REPORT WRITING

One of the most tedious tasks in conducting a full-blown RCA is writing the report. If no standard formats are available, this can be a laborious task that lacks continuity. Without standard formats, consistency in reporting results may suffer, leading to the possibility of the information being ignored or not understood.

In the manual method of writing reports, the analyst would generally use a word processing program and develop a stand-alone report with a table of contents that suits the team. Then some poor soul, usually the PA, is charged with the task of developing the content and typing it into an acceptable format. While the team members may contribute, the brunt of the legwork is on the shoulders of the PA. Then the task of properly distributing the report to the appropriate parties is at hand. All in all, this work is extremely burdensome and not the highlight of the analysis work.

PROACT provides the analysis team with the flexibility to customize their reports based on whom they are reporting to. As discussed in Chapter 10 earlier, a report going to TJC will look drastically different from one that might be used only internally. The software will allow the user to save report formats (table of contents)

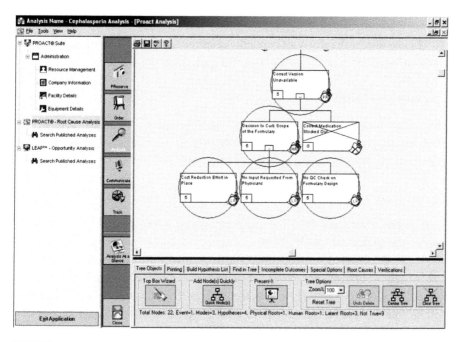

FIGURE 12.19 Adding root cause labels.

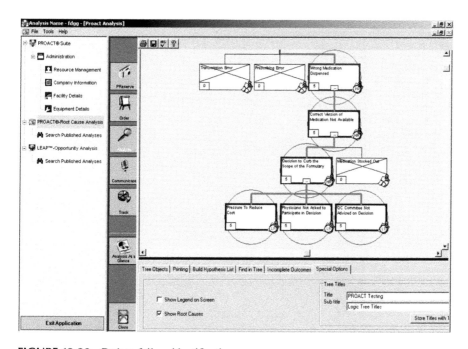

FIGURE 12.20 Path to failure identification.

for future use. This means the team could save a TJC-preferred format as well as an internal reporting format.

This section of the software basically has three tabs: (1) summary and findings, (2) recommendations, and (3) RCA report (Figure 12.21).

SUMMARY AND FINDINGS

This section is basically fill-in-the-blanks, where the analyst is asked to write a paragraph describing the event, a paragraph describing the findings, and a paragraph describing the RCA methodology used (whether PROACT or not).

RECOMMENDATIONS

Each block on the logic tree that has been labeled a root cause of any kind will prompt the analyst in this location to write an executive summary recommendation and a detailed recommendation. The analyst will also be asked to identify a metric to track, to name who will be responsible for the recommendation, to give an expected completion date, and to note the status of the recommendation. The analyst will be able to attach documents from other programs at this point as a file link if necessary. Sometimes, items such as new procedures will be attached to demonstrate that the recommendation has been completed and is ready to be implemented (Figure 12.22).

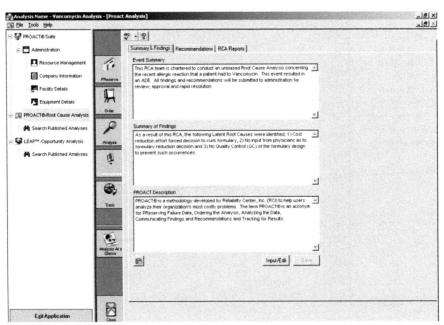

FIGURE 12.21 Communicate opening screen.

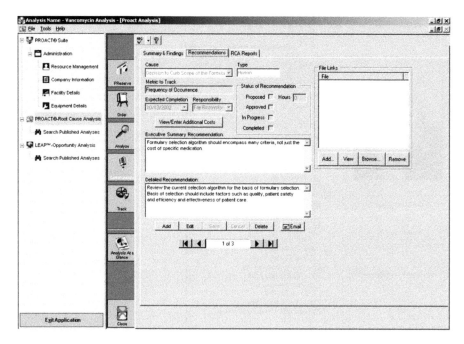

FIGURE 12.22   Recommendation tab.

RCA Report

This is where the components of the report are put together. The button labeled "esti-mated ROI" allows the analyst to assist in making the business case for approving the recommendations. Based on past losses and current expenses, basic ROI projec-tions can be made (Figure 12.23).

The "communicate report" button takes the analyst into a wizard, which asks the user to select the topics for the report, input the appropriate information for the report cover, and preview the output. Producing the report is as easy as that because it has been writing itself in the background as the analysis progressed (Figure 12.24).

Estimated Return On Investment Per Recommendation

Est. % of Annual Loss	Root Causes Identified	Annual Losses (Potential Benefit)	Sum of Recommendations	Estimated ROI (Year 1)
30%	Decision to Curb Scope of the Formulary	$150,000.00	$3,817.31	3929%
45%	No Input Requested From Physicians	$225,000.00	$12,048.08	1868%
25%	No QC Check on Formulary Design	$125,000.00	$2,163.46	5778%

Print...   Total Annual Loss: $500,000.00   Cancel   Save

FIGURE 12.23   Estimated ROI of recommendations sample.

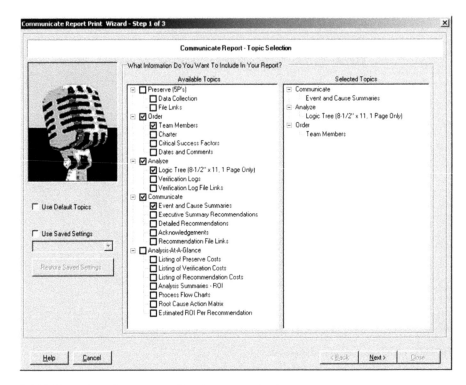

FIGURE 12.24   RCA report—table of contents selections.

AUTOMATING TRACKING METRICS

The team is not successful at RCA unless some bottom-line metric improves. Therefore, such metrics must be selected and monitored over time. In a manual format, analysts have to be diligent about getting certain data from certain reports, or they may have to develop a whole new report to get the information they seek.

The analyst should not make the tracking process so complicated that it is difficult and frustrating to accomplish. PROACT was designed to make this tracking process very simple, basic, and user-friendly. Tracking also has its own four-step wizard, which will walk the user through a series of prompts such as:

> Save graph as: _____
> Title of graph: _____
> Subtitle: _____
> Tracking intervals: _____
> Tracking periods: _____
> Tracking metric: _____
> Data to input: _____

This provides enough data to make an easy-to-follow basic graph. Each month, when new data is available, it can be input into the wizard to update the graph easily. This

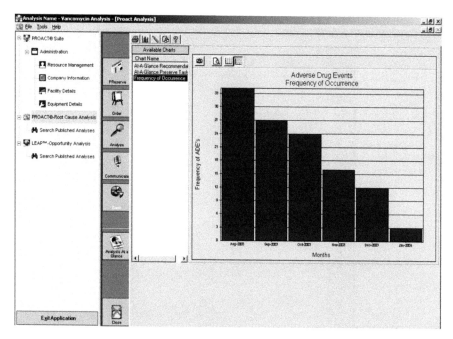

FIGURE 12.25   The tracking wizard—final chart.

data could be collected automatically if using PROACT in an enterprise environment that is integrated with an incident management system (IMS) (Figure 12.25).

If the analyst chooses to utilize more elaborate charting functions, it is easy to export the chart to a program such as Microsoft Excel and maintain a file link to its location.

The development of a dynamic tracking graph completes the circle of finalizing an RCA. Automating this graphing feature in PROACT alleviates the need to use a separate graphics package.

ANALYSIS AT A GLANCE

This feature will be invaluable to analysts who choose to use the business information features within PROACT. Analysts who have chosen to use features that encourage the recording of team members' time and expenses to conduct assigned tasks will find all of that information aggregated in this section. All costs associated with data collection, hypothesis verification, and recommendation development and implementation are totaled in this section under their respective tabs (Figure 12.26).

The "summary" tab is referred to as the dashboard for the PA, because with a single glance the PA can tell what percentage of the tasks assigned have been completed (or not completed). This section will also provide an estimated ROI (if the appropriate data has been input throughout the analysis) for the entire RCA. Earlier ROIs were projected for just the recommendations. Here the ROI is for the whole analysis, which includes costs associated not only with recommendations, but also with data collection and hypothesis verification (Figure 12.27).

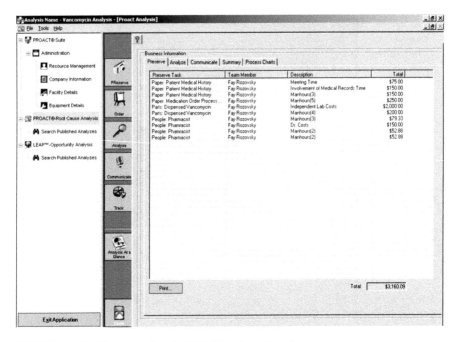

FIGURE 12.26   Costs associated with RCA tasks assigned.

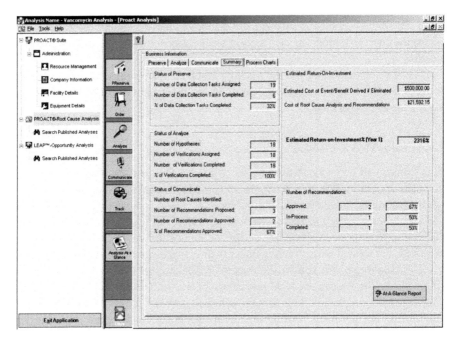

FIGURE 12.27   RCA "dashboard."

This information can be very beneficial to produce for the CFOs, who constantly want to see the returns for monies invested. Many CFOs may dispute or not see value in conducting these analyses, so this is good cost justification to present to such critics.

The last tab is for the process flow diagrams. When the analyst wants to use multiple tools to conduct the analysis, this block diagram feature may come in handy. If the reader will remember, this feature was also used in the OA and FMEA sections of the software. Had this RCA been created from those analyses, that chart would have automatically been imported into the RCA.

This tool allows the analyst to map out the process that failed and be able to describe, as a result of the RCA, the process as (1) it was designed, (2) it was when it failed, and (3) it will be per the proposed recommendation. These comparative process flow diagrams are excellent graphical tools to make key points to executive oversight committees (Figure 12.28).

PROACT KNOWLEDGE MANAGEMENT

Traditional manual methods require the use of a database package, a spreadsheet package, a word processing package, and a graphics package in order to complete every aspect of the RCA. This would require the alignment of file names and so on for continuity. PROACT compiles everything in one location, and the file can be e-mailed to others to assure proper distribution.

PROACT's enterprise version allows for efficient and effective knowledge transfer of the successful analyses to others in the system who may benefit. PROACT

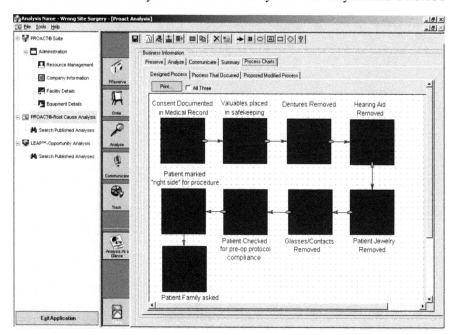

FIGURE 12.28 Comparative process flow diagrams.

instantly puts information at the fingertips of those who can use it most. Until an analysis is completed, only the PA and the team members can see their "works in progress." However, when the analysis is done, and after the PA and any other internal legal or regulatory personnel certify the analysis as complete, everyone who has permission and access to PROACT will be able to view the results. This feature is called *publishing*. Once an analysis has been published, its icon within the database changes allowing us to visually recognize those analyses completed. Only published analyses can be searched based on various criteria. If the organization does not want certain analyses to be published, it will simply not publish them, and they will remain confidential (Figure 12.29).

These features, as well as many others, allow analysts to focus more on doing the analysis rather than on the administrative tasks required to document the process and transfer the knowledge. PROACT is truly a proactive tool when conducting a RCA!

Client experience shows that utilizing PROACT to facilitate RCAs, FMEAs, and OAs has reduced the administrative time to complete them by approximately 50%. This means that, from a productivity standpoint, analysts can complete more analyses in a given time period if they automate their proactive analysis efforts.

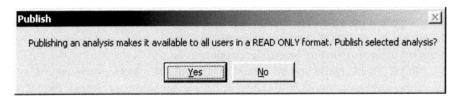

FIGURE 12.29   Knowledge manager—publishing.

VI

From Theory to Practical Application

13 Case Studies

CASE STUDY 1: FAILURE MODES AND EFFECTS ANALYSIS—INTERIM TRIAGE FMEA

Lead Analysts

Elizabeth Enz, BSN, RN, ARM, CPHRM, CPHQ
Stephen A. Craig, MHA

Purpose of the Study

The purpose of this study is to perform a prospective risk analysis for the interim emergency department (ED) walk-in entrance at Memorial Regional Medical Center (MRMC) using healthcare failure mode and effect analysis (HFMEA). Through the application of HFMEA, the team will be able to identify potential failures and subsequent solutions for the interim process. The solutions will be translated into detailed action plans that will be applied prior to opening the entrance.

Background

Bon Secours Memorial Regional Medical Center (MRMC) is a 225-licensed-bed facility. Experiencing the largest growth in the last five years is MRMC's ED.

The historical and projected growth has prompted the development and eventual construction of an expansion of the current ED, which will double the size of MRMC's existing ED (19 beds to 41 beds) (see Figure 13.1). The challenge became apparent to the team relative to finding a suitable walk-in entrance for an interim period. Reviewing the current footprint of the facility, the project team devised proposed alternatives.

One positive aspect of the triage relocation is that the change has afforded leadership the opportunity to implement best practices for ED triage. MRMC will now have dual triaging capabilities with two treatment rooms instead of one space in the current design. There will also be a physician at triage, which greatly improves efficiency. Because of implementation of these best practices and a redesigned process in a new location, potential failure modes, or areas where the process may break down, will potentially occur, precipitating the need for the team to perform an HFMEA.

The ED walk-in entrance relocation has triggered some genuine concerns of management, staff, and physicians in the ED. These concerns, which will be brought out when the team performs the HFMEA, include:

Facility	1999	2002	Bays	Vists/Bay
Chippenham Medical Center	46,732	61,447	31	1,982
Memorial Regional Medical Center	**24,232**	**36,750**	**19**	**1,934**
Richmond Community Hospital	13,333	22,738	13	1,749
Henrico Doctors' Hospital-Parham	13,886	17,978	12	1,498
Johnston-Willis Medical Center	29,990	36,549	26	1,406
Henrico Doctor's Hospital-Forest	26,634	32,018	23	1,392
Bon Secours- St. Mary's Hospital (Richmond)	41,299	42,595	31	1,374
Retreat Hospital	6,469	12,814	10	1,281
VCU Health System	82,979	80,921	80	1,012

Source: Viginia Health Information – 1999 & 2002 Annual Licensure Survey

FIGURE 13.1   Local market ED growth rates.

- Patient satisfaction
- Patient safety
- Public relations
- Staffing availability
- Physician satisfaction
- Patient/family interaction

STATEMENT OF THE PROBLEM

Given the current state of affairs at MRMC, and specifically the ED, the interim entrance presents a clear opportunity for potential failures throughout the emergency patient flow process, resulting in potential harm to patients. The consequences can be severe from a risk management standpoint as well as from a loss revenue standpoint.

A logical way to address these problems and managerial concerns is to convene a multidisciplinary team of professionals at MRMC to perform an HFMEA on the interim emergency triage process from the time a patient arrives on campus until he is taken back to the main ED treatment area.

HFMEA will ensure that each step in the emergency triage process will be examined on a systematic and prospective basis. The HFMEA process promotes systematic thinking about the patient care processes (i.e., what could go wrong, what needs to be done to prevent failures).

OBJECTIVES

Through the process of HFMEA, this management study will attempt to dissect the very detailed process of emergency patient entrance and triage and locate potential failures throughout process.

The objectives of this study are to identify as many potential failures and their corresponding modes or causes as possible and identify solutions to eliminate their occurrence or to mitigate their impact. The solutions will be in the form of detailed action plans that will be developed by the team members. In order to quantify the action plans in terms of how well the hospital is performing, the team will set measurable goals relative to each plan.

Patient satisfaction scores will be used as one indicator of performance. Previous scores from the ED will be compared to the first-quarter of scores once the area is open and operational. Another indicator of performance will be patient walkout rates and number of patients leaving against medical advice (AMA). Data will be compared to see what fluctuations, if any, have occurred as a result of the opening of the new entrance. Patient safety indicators will also be tracked.

To address concerns related to unanticipated events, provisions will be made to keep this analysis dynamic. Some of the improvement activities will consist of regular meetings with the HFMEA team to evaluate the process and review the proposed failures. This will make the HFMEA a true "knowledge management" tool.

METHODOLOGY

HFMEA studies are prospective in nature and are defined by the National Center for Patient Safety (NCPS) to be "a prospective assessment that identifies and improves steps in a process thereby reasonably ensuring a safe and clinically desirable outcome. HFMEA is a systematic approach to identify and prevent product and process problems *before they occur.*"* Leadership at MRMC recognized the value of this methodology and endeavored to put it in place before any major construction or related changes are to occur at the hospital.

HFMEA can be completed entirely on the computer, thus eliminating the manual paper process. To perform our HFMEA, the team was able to use software licensed to Bon Secours. The specific application used with the software program uses probabilistic data (basic failure modes and effects analysis) or historical data (opportunity analysis) to identify the "significant few"—those critical events that cause 80% of the failures.

Once the process was identified, the next step was to assemble a multidisciplinary team who will work together to perform the HFMEA (Figure 13.2).

Title
Administrative Resident–Team Leader
Director of EMS Services-MRMC
Director of Patient Access-MRMC
Senior Risk Manager-Bon Secours Richmond Health System
Director of Engineering and Maintenance
Director of Patient Advocate Services–MRMC
Medical Director of Emergency Department-MRMC
Emergency Department Nurse Manager–MRMC
Director of Safety and Security–MRMC
Contractor–HITT Construction
Administrative Director of Patient Care Services-MRMC
V.P. Reliability Center, Inc Software Company

FIGURE 13.2   HFMEA team member titles.

* NCPS, 2002.

The first task of the team was to flowchart the interim ED triage process using the LEAP software program (Figure 13.3). Each process step on the flowchart has a letter that will correspond to the HFMEA worksheet. The worksheet is where each subsystem under the process steps is analyzed using the HFMEA methodology of identifying potential failures (events) and modes (causes) and then assigning a probability and severity score. Determining severity and probability for each failure or event will be the result of the combined effort of the multidisciplinary HFMEA team (Figure 13.4).

This worksheet is able to provide the team with the blueprint of our efforts. Once the worksheet is complete and the team identified all of the potential failures, the software program calculated the "significant few," or the critical events that accounted for 80% of the risk of failure to one or more steps in the process. The graph and corresponding worksheet are shown in Figure 13.5.

The darker shaded bars represent the significant few. Those 17 items are then used to complete the final step in the process, which is the action plan.

The action plan will contain the following sections: item, discussion, action, responsible person and follow-up. The action plan will then have measurable indicators to gauge its efficacy.

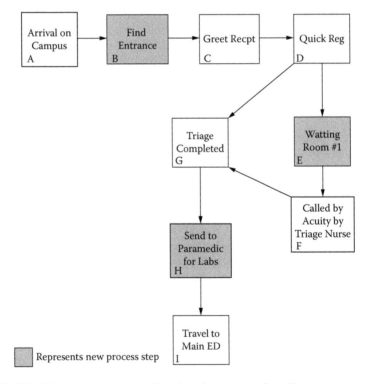

FIGURE 13.3 Emergency department interim triage process flow diagram.

Subsystem	Event	Mode	Probability	Severity	RPN
Waiting Room #1	Family Rejoins Pt in Waiting Room #1	Delay to Getting to Main ED	4	7	28
Travel to Main ED	Delay to Main ED	Multiple Travel Routes	4	7	28
Called By Acuity By Triage Nurse	Delay to Triage	Patient Backlog	3	7	21
Greet Reception	Adverse Pt Event	Lack of Staff for Transport	3	7	21
Travel to Main ED	Unable to Get to Main ED	Main ED at Capacity	3	7	21
Travel to Main ED	Deterioration of Pt Acuity	Travel Distance to Main ED	3	7	21
Greet Reception	Utilize the Wrong Entrance	Unfamiliar with Interim Entrance	3	6	18
Called By Acuity By Triage Nurse	Wrong Acuity Assignment	Vague Presentation from Pt	3	6	18
Travel to Main ED	Delay in Treatment	No Parking Available	3	6	18
Called By Acuity By Triage Nurse	Labs Re-Processed in Main ED	Nursing Competency Assessment	3	6	18
Find Walk-In Entrance	No Access to Entrance	Over-Crowding	4	4	16
Quick Registration	Wrong Pt Called Back	Lack of Organization	4	4	16
Travel to Main ED	Unable to Find Proper Entrance	Poor Signage	4	4	16
Greet Reception	Pt Dissatisfaction	Poor Signage	3	4	12
Greet Reception	No Receptionist Available	High Pt Volume	3	4	12
Called By Acuity By Triage Nurse	Wrong Acuity Assignment	Nursing Competency	3	4	12

FIGURE 13.4A HFMEA worksheet.

CONCLUSION

The interim triage process at Memorial Regional Medical Center (MRMC) began on 2/9/04 and was operational for approximately 11 months thereafter during construction. The work team identified 28 potential events that may result in an adverse patient event or outcome. Upon calculation, it was demonstrated that 17, or just fewer than 50%, of the events were within the "significant few." The "significant few"

Subsystem	Event	Mode	Probability	Severity	RPN
Triage Completed	No Protocol for Labs	Lack of Administrative Policies	2	7	14
Waiting Room #1	Pt Walks Out	Excessive Wait Times	3	4	12
Waiting Room #1	Pt Condition Worsens	Excessive Wait Time	3	4	12
Travel to Main ED	Leave the Campus	Campus Congestion	3	1	3
Greet Reception	Bypass Receptionist	Excessive Queueing at the Desk	3	4	12
Travel to Main ED	Death	Poor Signage	1	10	10
Quick Registration	Delay to Quick Register	Excessive Pt Volume	2	2	4
Quick Registration	Delay to Get Quick Registered	Not Captured at Rec. Desk	2	2	4
Quick Registration	Incorrect or Lack of Information	Poor Delivery of Info from Pt/ Family	2	2	4
Send to Paramedic for Labs	No Paramedic Available	Pt Transport	2	1	2
Called By Acuity By Triage Nurse	Improper Order to Main ED	Wrong Triage Classification	1	1	1

FIGURE 13.4B HFMEA worksheet.

represents those events that could potentially result in the most harm to the patient. A risk priority number (RPN) is calculated by multiplying the event's probability by the severity. The higher the RPN, the more likely the event is to happen with higher associated harm to the patient. In this case, 17 items were found to cause 80% of the risk associated with the interim triage process as revealed by the RPN value of 310, which is 80% of the overall RPN of 386. Our analysis showed that 50% of the potential failures would cause 80% of the total risk with this new process.

The results seem to show a significant amount of potential risks/failures in the process. "Modes" or reasons for failure, such as patient overcrowding and poor signage, were the two modes that seemed to be the cause of more than one potential failure.

This prospective analysis of the interim triage process was a great exercise for staff and physicians to participate in. It seemed to enlighten many of them and force them to analyze a process and its potential failures before an adverse event occurs. Through identification of the potential failures and potential harmful patient events, the team was able to develop specific action plans before the walk-in entrance became operational.

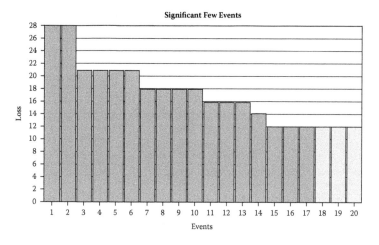

FIGURE 13.5 Significant few.

ID	Event	Mode	RPN
1	Family rejoins patient in wating room # 1	Delay to getting to main ED	28
2	Delay to main ED	Multiple travel routes	28
3	Delay to triage	Patient backlog	21
4	Adverse patient event	Lack of staff for patient transport	21
5	Unable to get to main ED	Main ED at capacity	21
6	Deterioration of patient acuity	Travel distance to main ED	21
7	Utilize the wrong entrance	Unfamiliar with the interim entrance	18

8	Wrong acuity assignment	vague presentation from patient	18
9	Delay to treatment	No parking available	18
10	Labs re-peocessed in main ED	Nursing coompetency assessment	18
11	No access to entrance	Overcrowding	16
12	Wrong patient called back	Lack of organization	16
13	Unable to find proper entrance	Poor signage	16
14	No protccel for labs	Lack of administrative policies	14
15	Patient dissatisfaction	Poor signage	12
16	No receptionist avibilable	High patient volume	12
17	Wrong acuity assignment	Nursing competency	12

RECOMMENDATIONS

Upon prospective review of the interim triage process, the work team recommended that specific action plans be created to address the events under the significant few. The action plans focused on the subsystems that presented the most potential risk to patients (Figure 13.6).

In order to determine the efficacy of the action plans developed, the team proposed a few measurable results that were tracked as the new process got under way. The first indicator, as stated in the objectives, is the patient satisfaction scores on five key indicators: overall quality of care, quality of nursing care, total time spent, likelihood of recommending friends/relatives, and overall teamwork between doctors, nurses, and staff. The team looked for scores that were statistically significant from the previous quarter. The results were an indicator of the effectiveness of the action plan. The team also tracked patient complaints as well as the patient walkout rate from the ED.

Relative to patient safety indicators, patient and visitor falls were also tracked.

To ensure that the measurable indicators mentioned above were regularly tracked and presented to management, the HFMEA work team routinely met to discuss the results and created a monthly report to senior leadership highlighting progress achieved relative to the action plan.

Item	Discussion	Action	Resp Person	FollowUp
Travel to main ED	Prospective failures to occur because of multiple travel routes to main ED.	Identification of proper route for patients and family members to use. Red plastic handrail covers to be used to signify proper travel route to main ED.	SC	Final route decided upon by management; proposal to be obtained for red handrail covers, approval required from EVP and CEO. Product to be delivered 2/20/04 and will take one day to install.
Travel to main ED	Prospective failures to occur because of lack of transport staff to main ED.	Plan A will be to hire five new transport staff to cover the area 24/7.	JR	New position proposal to be completed and sent for management approval.
		Plan B will be to utilize current transport staff and offer overtime shifts to cover the area.		Identify current PRN staff that would be available to fill in as needed or work extra shifts.
Travel to main ED	Prospective failures to occur because of travel distance to main ED.	Transport staff and volunteers to be utilized. Volunteers to be stationed near the exit of the triage area.	SC JR	SC to meet with volunteer director to establish guidelines for the new volunteer position and identify potential candidates.
		Establish position within volunteer corps that roams between the main ED and triage route to assist patients.		JR to work with transport dept. director to identify staff to cover the area.
Find the walk-in entrance Utilize the wrong entrance	Potential failure mode due to the patient being unfamiliar with the interim entrance on the MRMC campus.	Public relations and marketing staff to devise PR campaign around public awareness of the new traffic patterns and location of walk-in entrance.	SC JH	Public relations communication plan devised for entire project (see exhibit B). Signage and PR communication to be reviewed by hospital management. Local newspapers to be targeted two weeks before opening of interim entrance.
Find the walk-in entrance	Poor signage.	Map out key locations around the campus that are traditionally seen as major thoroughfares.	SC BZ	Sign company to plot locations on campus exterior and interior and deliver and set up the week of 1/26.
		Develop specific wording for the signs.	KJ JH	

FIGURE 13.6   HFMEA action plan: MRMC interim ED triage.

CASE STUDY 2: ROOT CAUSE ANALYSIS— INTER-HOSPITAL TRANSFER MEDICATION ERROR

LEAD ANALYSTS

Anonymity requested.

EVENT SUMMARY AND BACKGROUND

A 71-year-old was transferred from an acute care teaching hospital to a second acute care hospital for rehabilitation services post embolic stroke. Both hospitals were members of the same health care system. Six hours after admission to the second facility, the patient suffered an unexpected brain hemorrhage.

The event involved an intra-hospital transfer, which caused some serious concerns regarding patient management among both physicians and staff, since the type of handoffs experienced in this case were similar to those occurring on a daily basis in the health system. It was believed there were systematic flaws involved in the transfer of this patient and uncovering them would result in possible system-wide recommendations to prevent recurrence in the future, thus protecting patients from exposure to harm. Although there was no proven relationship to the occurring brain hemorrhage, there were many system failures identified in the transfer process that could have potentially contributed to the event.

RCA Team Charter

The charter was to identify the root causes of the unexpected brain hemorrhage in the 71-year-old intra-hospital transfer patient. This includes identifying deficiencies in, or lack of, management systems. Appropriate recommendations for identified root causes would be communicated to management for rapid approval and implementation.

RCA Methodology

The accepting transfer hospital used the PROACT root cause analysis methodology. This approach is depicted by the following steps of the analysis, which make up its name:

PReserve incident data
Order the analysis team and its members
Analyze the incident data with the team
Communicate findings and recommendations
Track for bottom-line improvement

Some of the data collection efforts involved developing strategies to collect some of the following information related to the incident, as shown in Figure 13.7. The RCA team consisted of the positions shown in Figure 13.8.

As the data were collected according to the strategies developed by the team, they then had to put the pieces of the puzzle together using the facts (data collected). At this point, the team used the PROACT logic tree tool to depict the cause-and-effect relationships that led up to the undesirable outcome (Figure 13.9).

As part of their RCA methodology, each hypothesis developed had to be proven or disproven using the evidence or data collected. Hypotheses proven to be false received an "X" and a "0" rating, indicating that evidence supports they are absolutely *not true*. Hypotheses proven to be true became facts and received a rating of "5", indicating that evidence supports that they are absolutely *true*. Any hypothesis with a rating between 0 and 5 indicates an uncertainty due to the lack of credible data available.

Figure 13.10 (a through c) demonstrates the verification methods used for each hypothesis and the resulting outcomes (if completed). Based on the identified physical, human, and latent root causes, an executive summary of recommendations was developed (Figure 13.11). This matrix summarizes the key root causes identified, their root cause type, their recommended solutions, their estimated completion date, and whether or not the recommendation implementation has been completed as of this writing.

Preserve Tasks for Analysis: MC-Anticoagulation

Category	Data	Strategy	Team Member	Date	Completed	Hours
Paper	Lab results from transferring hospital	Request from transferring hospital	Risk Mgmt Coord.	9/29/2003	Yes	1
Paper	medical records	Request record from transferring hospital Request accepting hospital record Request ambulance record Request intake record from CCRU	Risk Mgmt Coord.	9/26/2003	Yes	1
People	Chief Rehab	Interview re transfer to unit	Risk Mgmt Coord.	9/26/2003	Yes	2
People	RN manager CCRU	Interview re transfer to unit	Risk Mgmt Coord.	9/26/2003	Yes	2
People	Staff RN CCRU	Interview re transfer to unit	Risk Mgmt Coord.	10/10/2003	Yes	1
People	Case Mgr CCRU	Interview re transfer to unit	Risk Mgmt Coord.	10/10/2003	Yes	0.15
People	Pharmacist	Interview re transfer to unit	Risk Mgmt Coord.	10/2/2003	Yes	1
People	Intensivist	Interview re transfer to unit	Director RM/QA	9/30/2003	Yes	1
People	Risk Mgr transferring hospital	Interview re transfer to unit	Risk Mgmt Coord.	9/18/2003	Yes	1
People	RN Mgr CCU	Interview re transfer to unit	Risk Mgmt Coord.	10/10/2003	Yes	1
People	House Officer	Interview re transfer to unit	Risk Mgmt Coord.	10/10/2003	Yes	0.3
Paper	Transfer Procedures	Request from CCRU RN Mgr Risk Mgr transferring hospital	Risk Mgmt Coord.	10/26/2003	Yes	1
Paper	Anticoagulation Protocols	Request from accepting hospital CCRU RN Mgr Request from transferring hospital Risk Mgr	Risk Mgmt Coord.	10/6/2003	Yes	0.45

FIGURE 13.7   Sample of data collection tasks assigned.

Title
Risk Coordinator–Team Leader
Director of RM/QA
PhD
Quality Coordinator
Director of Pharmacy
M.D.–Chief of Medicine
Nurse Manager-Rehab
Chief of Rehab

FIGURE 13.8   RCA team member titles.

In conclusion, both facilities assumed that their Coumadin dosing schedules were the same. This was found not to be the case, and the patient who had received a 5-mg dosage of Coumadin prior to the transfer received another 7.5 mg dosage within 2.5 hours. A series of miscommunications and misinterpretations as to test results also contributed to this adverse outcome as evidenced in the logic tree (Figure 13.9) and associated verification logs (Figure 13.10).

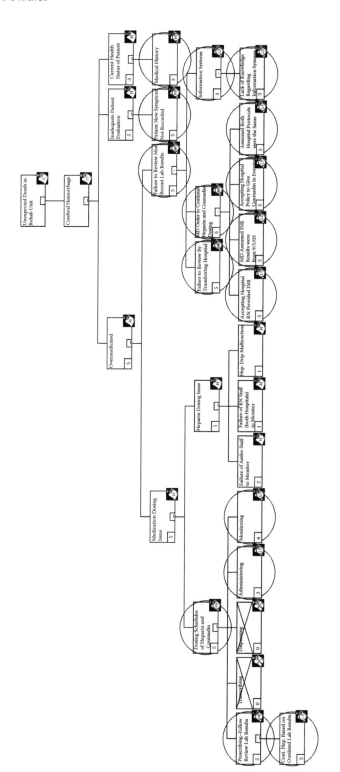

FIGURE 13.9 Brain hemorrhage logic tree.

Verification Logs

Hypothesis	Team Member	Verification Method	Outcome
Accepting hospital policy to give Coumadin in evening	Risk Mgmt Coord.	Review of accepting hospital's policy	Anticoagulation policy reviewed
Accepting hospital RN provided INR	Risk Mgmt Coord.	Review of transferring hospitals lab results transferred with patient	
Administration	Risk Mgmt Coord.	Review medication order sheet in medical record	
Administration	Risk Mgmt Coord.	Review physician order in medical record. Review medication order sheet in medical record	MD order for 7.5 mg coumadin and 800 u heparin. INR >2.0 to discontinue.
Assumed both hospitals, protocols were the same	Risk Mgmt Coord.	Review of accepting hospital's policy and Review of transferring hospital's policy	Review of policy substantiated Coumadin given in evening at accepting hospital
Cont.Hep. based on outdated lab results	Risk Mgmt Coord.	Review of transferring hospital's lab results transferred with patient	
Current health status of patient	Risk Mgmt Coord.	Review of medical history Existing co-morbidity	Patient risk factors of A Fib, HTN, previous TIA, syncope, 60-80% right carotid stenosis
Failure of ambo staff to monitor	Risk Mgmt Coord.	Review ambo transfer record	

Hypothesis	Team Member	Verification Method	Outcome
Failure of RN staff (both hospitals) to monitor	Director RM/QA	Interview nursing staff	Agency nurse indicated heparin infusing according to order
Failure of RN staff (both hospitals) to monitor	Risk Mgmt Coord.	Review medical records	Review of medical records verified that heparin dosing according to order
Failure to review by transferring hospital	Risk Mgmt Coord.	Review 9/5 lab order	Lab result identifies there was "urgent" test drawn on 9/5/03 at 12 PM
Failure to review by transferring hospital	Risk Mgmt Coord.	Review 9/5 lab order	Lab result identifies there was "urgent" test drawn on 9/5/03 at 12 p.m.
Hep. drip malfunction	Risk Mgmt Coord.	Unable to determine	
Heparin dosing	Risk Mgmt Coord.	Review transferring hospital's "Transfer Summary"	Transfer summary indicates most recent INR 1.5 Failure by accepting hospital to confirm INR of 1.5 was 9/5 result
Information systems	Risk Mgmt Coord.	SMS system - ability to access reports from other hospital	Ability to access lab reports on pts transferred from other intr-system hospitals
MD assumed INR results from 9/5/03	Director RM/QA	Review medical records	INR results in transfer summary from 9/4/03
Medical history	Risk Mgmt Coord.	Review transfer summary and transferring hospital's medical records	Medical record confirmed this hypothesis

FIGURE 13.10 Verification logs. (*Journal of Healthcare Risk Management,* American Society for Healthcare Risk Management (ASHRM), 24(3), 2004. Used with permission.)

Hypothesis	Team Member	Verification Method	Outcome
Medication dosing issue	Risk Mgmt Coord.	MAR from transferring hospital MAR from accepting hospital Pyxis records from transferring and accepting hospitals Lit search regarding heparin and coumadin	Evidence supports patient was indeed over medicated.
Monitoring	Risk Mgmt Coord.	Medical records	5mg coumadin given at 5 pm 9/5/03 at transferring hospital 7.5 mg coumadin given at 7:30 pm on 9/5/03 at accepting hospital
Over medicated	Director RM/QA	INR results PTT results	INR 2.9 on 9/5/03 drawn at 12:28 PM PTT 80.2 on 9/5/03 PT 25.1 on 9/5/03
Patient new symptom	Director RM/QA	Review medical record Interview staff	Documentation in medical record supports that pt evaluation was not adequate. Patient c/o headache for 24 hours prior to transfer to CCRU and continued to complain after admission. Transferring hospital "Nursing Transfer Summary" done during the day and not updated prior to transfer.
Prescribing - follow review of lab results	Director RM/QA	Interview with prescribing physician	Prescribing physician was going by the lab results from the transfer summary dictated the morning prior to pt arrival

FIGURE 13.10 (continued)

Executive Summary Recommendations

Root Cause	Type	Recommendation	Responsible	Estimated Completion Date	Completed
Failure to review most recent lab results	Human	Develop interfintra transfer process.	Director RM/QA	11/20/2003	No
Monitoring	Latent	Patients receiving IV anticoagulation will no longer be candidate for admission to rehab unit.	Director RM/QA	10/1/2003	Yes
Information Systems	Latent	Nursing Staff Development will provide staff education on SMS. Chief of Staff will communication to physicians.	Director RM/QA	11/4/2003	Yes

FIGURE 13.11 RCA executive summary recommendations.

CASE STUDY 3: SPECIMEN INTEGRITY OPPORTUNITY ANALYSIS

LEAD ANALYSTS

Anonymity requested.

STATEMENT OF THE PROBLEM

A 225-bed acute care facility observed an increase in the costs associated with their blood drawing systems in their ED. In an effort to quantify the severity and magnitude of the potential problem, they commissioned a team to conduct an opportunity analysis (OA) to review their blood drawing process and determine the scope of the problem.

OBJECTIVES

The objective of the analysis was to identify the significant few events in the process. These are the 20% or less of the events that are causing 80% or more of the physical losses in the system being analyzed. This analysis will quantify and then prioritize the annual impacts of each of these events and their respective modes, allowing management to make educated decisions about appropriate corrective actions.

METHODOLOGY USED

The analysis team utilized the OA process described in detail in Chapter 5 of this text. Figure 13.12 is the process flow diagram (PFD) used to define the scope of the process to be analyzed (where it begins and where it ends). The impacts/occurrence used were man-hours, materials (supplies), and lost profit opportunities (instances where the hospital had to pay for errors when otherwise they could have been making money). The following assumptions were used to derive the impacts per occurrence. These assumptions were derived from data provided by the accounting department.

ASSUMPTIONS USED

The impacts used in this analysis are meant to include the following direct and indirect costs associated with a need to redraw blood due to an error on the first attempt.

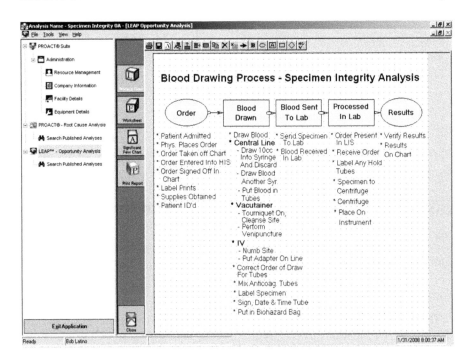

FIGURE 13.12   Specimen integrity process flow diagram.

1. Additional labor costs: Average costs per redraw for:
   a. RN = $2.62 (7 minutes @ $22.50 per hour)
   b. Unit secretary = $0.67 (5 minutes @ $8.00 per hour)
   c. Lab tech = $1.11 (5 minutes @ $13.30 per hour)
   d. Med tech = $2.31 (7 minutes @ $19.84 per hour)
   e. QA = $1.40 (3 minutes @ $28.00 per hour)
2. Additional material costs: Average costs for supplies associated with redraw:
   Venipuncture supplies
   a. Tubes × 3 per SST, lavender, citrate = $0.28 each
   b. Tourniquet = $0.11 each
   c. Butterfly = $0.75 each
   d. Luer adapters = $0.15 each
   e. Gauze, alcohol, and band-aid = $0.10 each
   Lab testing supplies
   a. Reagents = $1.00 per test ran, 33% ran
3. Lost profit opportunities costs associated with using valuable ED time and space to redraw a blood sample after a failed attempt

It was estimated that the average redraw in the ED would cause the patient to remain in the ED an additional 30 minutes (7 minutes for the redraw, 20 minutes to test the sample, and 3 minutes for routing the sample to and from the lab). Had this redraw not occurred, the space in the ED could have been used for waiting patients, where the time is billable.

The ED accrues by points. The average revenue per unit in the ED is $828. The average length of stay (LOS) in the ED is 3 hours 45 minutes, or 225 minutes. Therefore, the average cost per minute in the ED is $3.68.

A literature search reveals on the conservative side that the average impact of all costs (direct and indirect) associated with a blood culture contamination is $5000.

The team acknowledges that these reflect the physical costs associated with such events and *do not* reflect the impacts associated with additional risk or customer dissatisfaction. It is acknowledged that such esoteric parameters are difficult to measure in dollars but could easily outweigh the dollar values expressed.

Figure 13.13 is the opportunity analysis spreadsheet that was developed by the analysis team.

CONCLUSIONS

This analysis demonstrated that 4.8% of the occurrences (480/10,013) are causing 82% of the annual dollar losses ($2,400,000/$2,896,560). At the time of this analysis there were approximately 10,013 redraws per year (extrapolated from the period of 9/03 to 9/04) resulting in a consumption of man-hour, material, and lost profit opportunities of $2,896,560.

Figure 13.14 demonstrates the distribution of losses for each type of event experienced. Clearly, "blood contamination" is the single most significant mode contributing to the need for redraws.

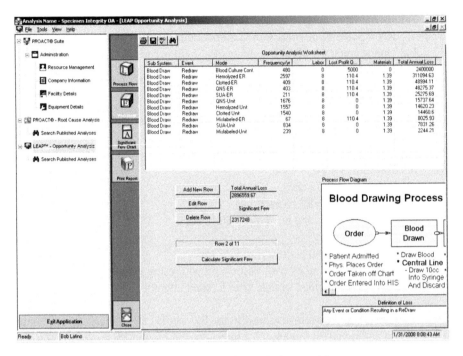

FIGURE 13.13   Specimen integrity opportunity analysis spreadsheet.

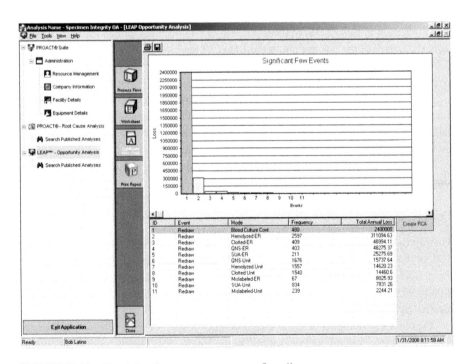

FIGURE 13.14   Blood drawing process—process flow diagram.

RECOMMENDATIONS

A literature search reveals that the use of a well-trained phlebotomist staff will result in 98% successful draws on the first attempt. Given that statistic, this would indicate that a savings of $2,838,629 ($2,896,560 × 0.98) would be realized under the current conditions. The cost of a phlebotomist staff of 25 (full-time equivalents) is estimated at $697,400 per year.

The cost/benefit then becomes:

Total potential returns—$2,838,629 per year
Total initial investment—$ 697,400 per year
Potential return on investment (ROI)—407%
Estimated payback period— ~3 months

Based on this empirical data, it is recommended that the facility set up a phlebotomy team in house to bring consistency to the task of drawing blood, reduce the risk of exposure due to errors currently being caused in the blood drawing process, and increase overall patient safety.

Appendix

Paradigm Shift Summary

The following is a synopsis of the paradigm shifts discussed in this text that are necessary to move a reactive culture toward a proactive one.

#	Current Paradigms	New Paradigms
1	Patients rely solely on their caregivers to protect them.	Patients must accept responsibility for their own safety in a hospital and realize they are a part of the care process.
2	Risk managers typically are chartered to be prepared for adverse outcomes in terms of adequate insurance coverage and handling claims after the fact.	Risk managers should be chartered to identify and quantify the risks associated with their organizations and take measures to reduce the risk of the more critical outcomes before they occur.
3	Healthcare today must be safe, as we are using all of this fancy technology.	Medical diagnostic technologies represent the state of the art in the world today. However, the technology employed to help caregivers effectively communicate with each other to ensure patient safety is poor at best. Administrative technologies must be implemented to provide caregivers the most updated information they need—instantly!
4	Root cause analysis (RCA) is something that engineers use in industry to solve complex and costly problems.	Root cause analysis (RCA) is specific to the human being and how we use our brains to solve *any* undesirable outcomes.
5	We do RCA in order to comply with regulatory requirements.	We do RCA to increase patient safety— bottom line!
6	Undesirable outcomes occur because of one cause.	On average, we must understand that it takes 10 to 14 cause-and-effect relationships to queue up in a particular pattern for an undesirable outcome to occur.
7	Brainstorming is RCA.	True RCA involves the identification of physical, human, and latent root causes fully supported with hard evidence.
8	True RCAs take too much time to complete, and we do not have that much time.	If an RCA technique being used is not evidence based, it is not RCA. If true RCAs are not being completed, the system runs the risk of recurrence of the event and the consequences that accompany it.

#	Current Paradigms	New Paradigms
9	FMEA is the same as RCA	FMEA analyzes a process/system. RCA analyzes an individual event
10	RCA and Six Sigma are the same thing, so why do we have to do both.	RCA and FMEA are subsets of Six Sigma that can aid in project selection and identification of true root causes. RCA and FMEA can significantly decrease man-hours to complete projects, thus reducing the cycle times of projects and realizing ROIs much sooner. RCA and Six Sigma are tools that will reduce the reactivity of the organization and ease the burden on the overworked staff.
11	RCA and Six Sigma approaches are just more "flavor of the month" types of things that executives think up and we have to waste our time on.	Tools such as RCA and Six Sigma are being provided to us to help reduce the reactive nature of our environment. By making the environment more proactive, we free up the time to do what is really important, and we reduce our stress level, which degrades our quality of life. Most importantly, we increase patient safety, which is the primary focus and the reason for our existence!
12	Failures typically occur due to one root cause.	On average, it takes a series of cause-and-effect relationships to queue up, in a particular sequence, for an adverse outcome to occur.
13	Healthcare is a complex system, and therefore only complex solutions can solve their problems.	Healthcare is like any other complex system where the system reliability is dependent upon the effective communication and decision-making skills of the people who work within the system.
14	RCA is viewed as another task or a burden to the analyst.	RCA is viewed as a tool to relieve the current burdens on the analyst.
15	RCA is only a reactive tool.	RCA can be applied proactively as well as reactively.
16	Hearsay is treated as fact in brainstorming techniques.	Hearsay is treated as assumption in true RCA techniques.
17	Data collection is a luxury to an investigation.	Data collection is a necessity to an RCA.
18	RCA outcomes (causes) are subject to approval of the executive management.	RCA outcomes (when done properly) are nonnegotiable—they occurred, and it is a fact!
19	Proactive tools are non-value added to the organization.	Proactive tools are fiscally responsible tools and create value to the organization.
20	Effective communications are impossible in a hospital because of the complexity of the system.	IT solutions exist today in other complex systems to quantumly improve communications within the system.

Index

A

Accreditation agencies, 9
ADE, *see* Adverse drug event
Adverse drug event (ADE), 45, 52
Against medical advice (AMA), 173
AMA, *see* Against medical advice
Analysis team, ordering of, 89–98
  cause-and-effect relationship, 97
  challenges of RCA facilitation, 92–94
    acceptance of opinions as facts, 93
    arguing among team members, 94
    bypassing of RCA discipline and going
      straight to solution, 93
    dominating team members, 93
    floundering of team members, 93
    going off on tangents, 93–94
    reluctant team members, 93
  codes of conduct, 95
  CSF examples, 96
  difference between team and group, 90
  error-change relationship, 97, 100
  key performance indicators, 95
  mission statement, 95
  principal analyst characteristics, 92
  promotion of listening skills, 94–95
    don't interrupt, 94
    one person speaks at a time, 94
    react to ideas, not people, 94
    separate facts from conventional wisdom,
      95
  team charter, 95
  team codes of conduct, 95
  team critical success factors, 95–96
  team meeting schedules, 96–98
  team member roles and responsibilities,
    90–92
    associate analyst, 91
    critics, 91–92
    experts, 91
    principal analyst, 90–91
    vendors, 91
  three-knock rule, 95

B

Black Belt, 66–69, *see also* Six Sigma
  data display tools, 67
  Internet polls, 66
  lean thinking tools, 68
  problem-solving tools, 67
  process improvement tools, 68
  process mapping tools, 67
  product/process interaction tools, 68
  root cause analysis tools, 67
  statistical tools, 67
Brainstorming
  data collection, 89, 98
  data needs, 84
  interrogation by, 49
  paradigm shift, 189
  Six Sigma and, 68
  techniques, 62
Broad and all-inclusive thinking, 105, 106

C

Career-limiting activities, 26
Case studies, 171–187
  failure modes and effects analysis (interim
    triage FMEA), 171–177
    background, 171–172
    conclusion, 175–176
    lead analysts, 171
    methodology, 173–174
    objectives, 172–173
    purpose of study, 171
    recommendations, 177
    statement of problem, 172
  HFMEA methodology of identifying
    potential failures, 174
  knowledge management tool, HFMEA as,
    173
  LEAP software program, 174
  root cause analysis (inter-hospital transfer
    medication error), 178–183
    event summary and background, 178–179
    lead analysts, 178
    RCA methodology, 179–183
    RCA team charter, 179
  specimen integrity opportunity analysis,
    183–187
    assumptions used, 184–185
    conclusions, 185
    lead analysts, 183
    methodology used, 184
    objectives, 184
    recommendations, 187
    spreadsheet, 186
    statement of problem, 183
  verification logs, 182–183
Chronic events, 27, 28
Clinical journals, case histories in, 139
CMSs, *see* Computerized management systems
COLAs, *see* Cost of living allowances

For Product Safety Concerns and Information please contact our EU
representative GPSR@taylorandfrancis.com
Taylor & Francis Verlag GmbH, Kaufingerstraße 24, 80331 München, Germany

www.ingramcontent.com/pod-product-compliance
Ingram Content Group UK Ltd.
Pitfield, Milton Keynes, MK11 3LW, UK
UKHW021613240425
457818UK00018B/528